CSCPRC REPORT NO. 6

Wheat in the People's Republic of China

A Trip Report of the American Wheat Studies Delegation

Edited by VIRGIL A. JOHNSON and HALSEY L. BEEMER, JR.

Submitted to the Committee on Scholarly Communication with the People's Republic of China

NATIONAL ACADEMY OF SCIENCES
Washington, D.C. 1977

NOTICE: The views expressed in this report are those of the members of the Wheat Studies Delegation and are in no way the official views of the Committee on Scholarly Communication with the People's Republic of China or its sponsoring organizations--the American Council of Learned Societies, the National Academy of Sciences, and the Social Science Research Council.

The exchange visit of the Wheat Studies Delegation to the People's Republic of China was supported by a grant from the National Science Foundation. This visit was part of the exchange program operated by the Committee on Scholarly Communication with the People's Republic of China, founded jointly in 1966 by the American Council of Learned Societies, the National Academy of Sciences, and the Social Science Research Council. Sources of funding for the Committee include the Department of State and the Ford Foundation.

The Committee represents American scholars in the natural, medical, and social sciences, as well as the humanities. It advises individuals and institutions on means of communicating with their Chinese colleagues, on China's international scholarly activities, and on the state of China's scientific and scholarly pursuits. Members of the Committee are scholars from a broad range of fields, including China studies.

Administrative offices of the Committee are located at the National Academy of Sciences, Washington, D.C.

Library of Congress Catalog Card Number 77-90510

International Standard Book Number 0-309-02637-7

Available from

Printing and Publishing Office
National Academy of Sciences
2101 Constitution Avenue, N.W.
Washington, D.C. 20418

Printed in the United States of America

PREFACE

The U.S. Wheat Studies Delegation entered the People's Republic of China at Peking on May 19, 1976. The Chinese Association of Agriculture was its official host in China. During the ensuing 4 weeks the delegation traveled an estimated 7,000 miles through the wheat areas of central and north China. It visited agricultural research institutes, academies and colleges, communes and their subunits, grain storage facilities, a major wheat flour mill, and a bakery. The delegation also visited agricultural and industrial exhibitions in Shanghai, the Hai River control exhibit in Shihchiachuang, and historical sites in or near the cities visited.

The Wheat Studies Delegation was comprised of 11 U.S. scientists and scholars representing agronomy, plant breeding, genetics, soil science, cereal chemistry, plant pathology, plant physiology, agricultural economics, and Chinese history. A U.S. Foreign Agriculture Service agriculturalist stationed in Peking traveled with the delegation in China. Delegation objectives were (1) to study Chinese wheat research and its organization, (2) to examine wheat production and distribution methods, (3) to study processing of wheat and procedures employed to insure desired quality and nutritional standards, (4) to learn about management of soil, use of fertilizers, and irrigation practices, (5) to examine the economics of wheat production, processing, and distribution, and (6) to observe Chinese wheat germ plasm and its management and explore opportunities for the exchange of wheat germ plasm between the United States and China and Chinese participation in international wheat evaluation networks.

The delegation brought with it to China several sets of wheat germ plasm and selected books and reprints of scientific articles on wheat authored by delegation members. The germ plasm was presented to the Chinese Association of Agriculture for in-country distribution, and the printed materials were given to colleges, academies, research institutes, and communes visited by the delegation. In turn, the delegation received from the Chinese three sets of wheat and triticale germ plasm and copies of recent issues of Chinese scientific journals on agriculture and wheat. Lectures were presented by invitation at several institutes and colleges by delegation members.

Wheat is grown widely and intensively in central and north China. We were impressed by the broad expanses of wheat across Anhwei, Honan,

U.S. Wheat Studies Delegation members. Front row kneeling: R. H. Busch and R. H. Myers. Back row: D. N. Moss, W. E. Kronstad, A. P. Roelfs, R. A. Olson, R. J. Cook, V. A. Johnson, L. E. Eastman, Y. Pomeranz, and P. Schran (delegation member K. Neeley not in photo).

Welcome banner for the Wheat Studies Delegation at the Ma Lu Commune near Shanghai.

Shensi, Shansi, and Hopei provinces, where it occupied at least 60 percent and probably as much as 80 percent of the cultivated land. Many of the fields were 25-30 ha in size, some larger. Irrigated wheat predominated in the areas traveled. Most of the wheat we saw was good, probably because we were shown almost exclusively irrigated wheat but also because of very intensive and effective land-leveling and soil improvement programs and because of wide use of productive, disease-resistant varieties. Many of the irrigated wheat fields of north China had the appearance of irrigated wheat in the Snake River Plains and Columbia River irrigation districts of the U.S. Pacific Northwest.

Our Chinese hosts showed us mainly their best wheat and their most advanced production practices. We were shown only one nonirrigated wheat field; this exception was the result of repeated requests by the delegation to observe rain-fed wheat. The observed field was on fertile soil and had received ample precipitation. Its yield potential appeared to be comparable to that of adjacent irrigated fields. It was not a typical rain-fed field as judged by wheat we saw from a moving train between Sian and Peking. During the train ride we saw many wheat fields that were very poor, so poor in fact that it seemed doubtful that they would yield as much as 1,000 kg/ha. There is little likelihood that the expanse of poor wheat along 300-400 miles of tracks was due to diseases or insect pests. The problem seemed to be an inadequate supply of water, inadequate soil fertility, or possibly salt accumulation in the soil. To have observed the "best wheat technology" is a useful point of reference, but not seeing the poorer wheat of China, except from a moving train, we were unable to identify with certainty their more important wheat production problems.

The communes visited by our delegation probably were substantially above the ordinary. One of these, the Chiang Ning Commune, had been a part of the model Chiang Ning Hsien (county) during the 1930's. The Ma Lu Commune, located near a major city, was especially prosperous. A large exhibition hall and electric display suggested to us that Ma Lu is visited as a model commune by many other Chinese.

Timing of the delegation's travel in China was ideal for useful field observations of wheat production and research. The wheat was headed in all areas visited except the spring wheat area of northeast China. It was approaching maturity in the Shanghai and Nanking areas, but none had been harvested. The stage of development was near optimal, except in the northeast, for observation and assessment of most disease, insect, and other production problems.

The period of travel by the Wheat Studies Delegation in China was perhaps unique. Radicalism was at high tide. Chou En-Lai had died in January, and the campaign against Teng Hsiao p'ing, which had begun in March, was perhaps at its most strident level. Because of the high political tenseness during the delegation visit, there probably was even less indication of diversity of opinion than was noted by other U.S. delegations visiting China during periods of relative political calm.

Nonetheless, delegation members were deeply impressed with Chinese agriculture and the apparent progress of China in meeting its challenge of more food production. This report reflects these impressions. Delegation members provided major inputs to sections of the report as follows: "Soil and Crop Management and Related Factors," R. A. Olson and D. N. Moss; "Wheat Breeding and Genetics," W. E. Kronstad and R. H. Busch; "Pests of Wheat," R. J. Cook and A. P. Roelfs; "Wheat Quality, Storage, and Processing," Y. Pomeranz; "Politics and Populism: Implications for Wheat Production, Research, and Extension," L. E. Eastman; "Educational Programs and Manpower Training" (based on notes of delegation members), H. L. Beemer (CSCPRC Staff Officer); "Wheat Production and Distribution," R. H. Myers; and "Farm Labor and Living in China," P. Schran.

The delegation was accompanied throughout China by Mr. Li Teng-ch'un, wheat breeder and representative of the Chinese Academy of Agricultural and Forestry Sciences; Mr. Huang Yung-ning, secretary, Chinese Association of Agriculture; Ms. Yang Hung, interpreter, Chinese Association of Agriculture; and Mr. Huang Hao-hsin, interpreter, Chinese Association of Agriculture. These people contributed much to the success of the trip by their highly efficient and pleasant management of all travel details, including numerous special requests from the delegation.

V. A. Johnson

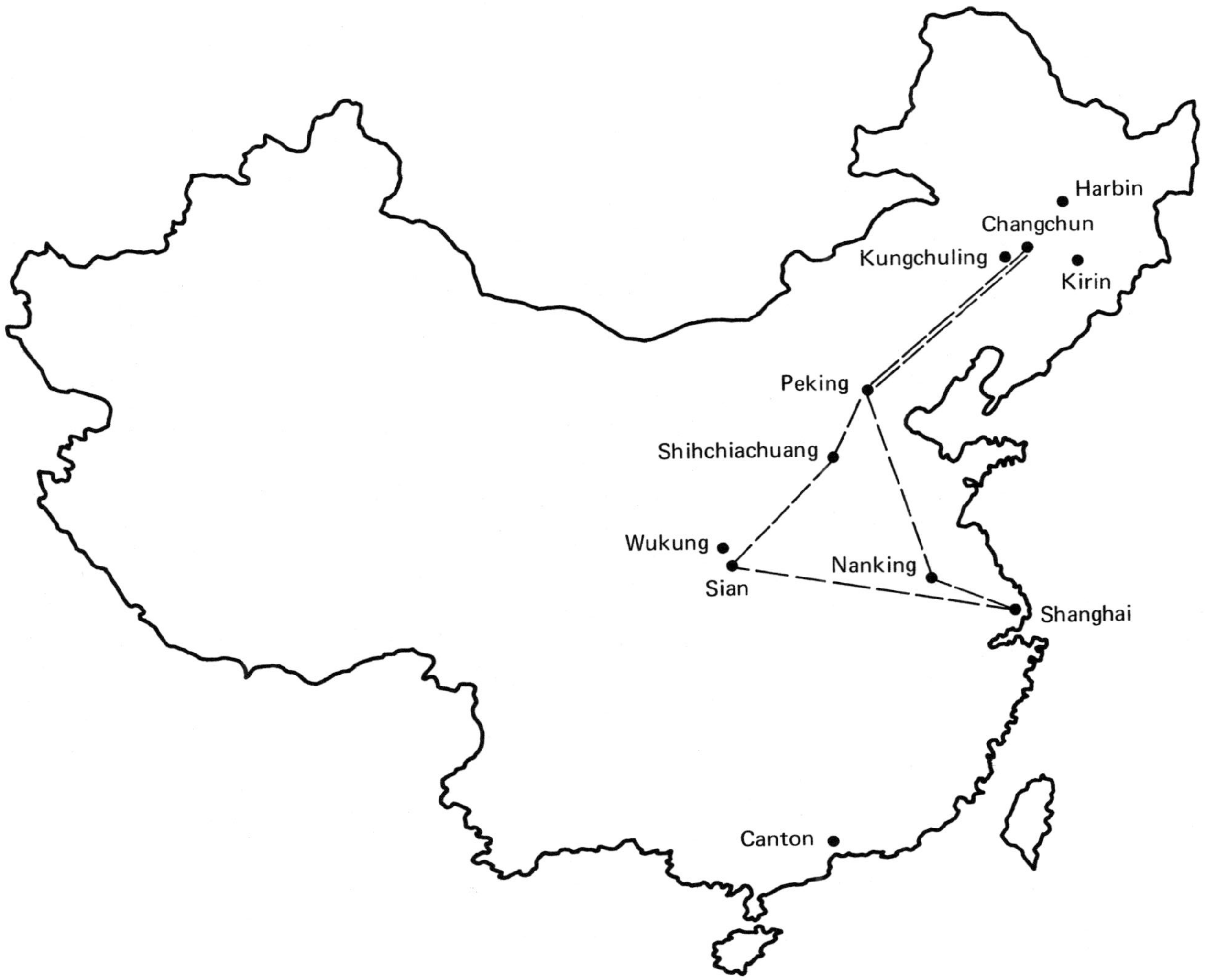
Harbin
Changchun
Kungchuling
Kirin
Peking
Shihchiachuang
Wukung
Nanking
Sian
Shanghai
Canton

CONTENTS

1

SOIL AND CROP MANAGEMENT AND RELATED FACTORS

SOIL CLASSIFICATION AND CAPABILITIES EVALUATION

The People's Republic of China has an extraordinarily wide range of soil conditions expressing the diversity in climatic and ecological features of the country. The original classification scheme effected by James Thorp of the United States and associates of the Soil Science Institute in Nanking during the 1930's remains as the authoritative one. Very little change appears to have been made in this essentially genetic approach to soil classification other than a few minor name changes at the soil group level. For example, the Shantung Brown soils of the original scheme are now called Cinnamon soils.

We were informed of a Soil Geography Division of the above institute, which is updating the basic classification of soils throughout the country. In our visits in the country, however, we never encountered a soil scientist versed in soil morphology and classification. Rather, at the commune level, such morphological classification of soil as existed was on the basis of field boundaries with descriptive terms like silt soil, sandy soil, clay soil, black soil, etc. Very often the issue was passed over with the contention that the peasant farmer from his experience in farming the land knew everything about the soil that was necessary. The Chinese attitude is in direct contrast with that of the Russians, who put their primary emphasis in the soil area on soil genesis and classification, with much less effort directed toward effective soil management and use.

A seeming misnomer in the classification system presently expressed is in the delineation of "loess soils." There is very evident loess cap on a great deal of the high benchlands and upland plains in the region of brown forest soils for which no loess connotation is given. The impact of loess deposit is very evident throughout much of the northeastern third of China, the locus of our entire visit, a most fortunate geologic phenomenon for the agricultural potential of the country.

A second answer given to us on the question of soil classification and mapping dealt with maps prepared from soil-testing data. In at least some counties, surface soil samples are collected from most fields every second or third year and subjected to quick tests for soil N, P, and K. From these data, charts are drawn to indicate the distribution of varied levels of these primary nutrients in soils of the country.

In one respect the detailed kind of soil classification and mapping carried out in the United States would not be apropos for the soils of China. The majority of the more intensely farmed soils have been so thoroughly modified by the continuous stirring over thousands of years and recent land-grading operations as to qualify essentially as man-made soils. There is probably no other country where modification of the textural profile of soil is undertaken, by the addition of sand to intractable clay soils to produce a more friable medium in one case and the addition of silt and clay to sandy soils to enhance the water and nutrient storage capacity of the soil in the other case. Only in a labor intensive country like China could such an operation prove feasible. One can only admire the high respect the population of the country obviously holds for its soil resource.

A major lesson taught in the "learning from Tachai," a nationwide campaign of emulating a model commune in north China, is the means for creating the "sponge soil." In essence, the message is to first stabilize the soil by benching or level terracing so that accelerated erosion is essentially eliminated. Subsequent to this land shaping, a regime of extensive organic manuring is followed, accompanied by deep plowing for ameliorating soil structure. The objective is a bulk density of 1 or slightly less in medium- to fine-textured soils for assuring a high permeability rate for the summer's rainfall.

The wheat study team unquestionably obtained a rather distorted perspective of soil quality from the sites visited relative to the entire country. With but one exception the communes studied were located on benchlands of the valleys or coastal plains where soils are formed on alluvial or loessial materials. Such materials are invariably associated with the most productive soils of any climatic/physiographic region. In the daylight portion of our train trip through the country, however, we observed wheat as the predominant crop on soils of much lower quality. This was true in some of the sharply sloping land of central Shensi and Shansi provinces as well as on very sandy deposits of the Hai River and on saline and alkali spots of this and other valleys. Nutrient deficiencies were evident in the wheat growing on many of these areas, and moisture stress was often apparent even on fields with an existing irrigation regime. From these cursory observations we might conclude that the Chinese are quite effectively distributing the limited water and nutrient resources available at this time on the better soils of highest potential productivity. No doubt, as more nutrients and energy become available and additional river harnessing is effected, these lower-quality soils will be brought into the more intensive agricultural production scheme. In essence, much has been accomplished, but as our hosts often expressed, a great deal remains to be done for realizing the full potential of the country's soil resources.

The above statement should not be construed to mean that good wheat growth was observed only on show places selected by our Chinese hosts. On the contrary, we saw a great deal of very good wheat in our automobile trips through the countryside. We have no basis for doubting that the country will indeed produce a bumper crop this year in comparison with prior years.

SOIL WATER MANAGEMENT

Irrigation

Repeated emphasis has been made in our discussions on the efforts toward comprehensive water control by the People's Republic of China since Liberation in 1949, serving the objective of weather control by water conservancy projects. Our observations confirm that tremendous strides have been made in irrigation development in particular. It has been claimed that some 80-85 percent of the wheat produced in China is irrigated and no longer dependent on the uncertainties of rainfall amount from year to year. We were advised that major river control projects for alleviating flooding and for storing surface water have contributed in a major way to this extension of the irrigated area and observed a number of new lined irrigation supply ditches for which the water source had to be rivers or large reservoirs (e.g., the newly created diversion of 100 m^3/s from the Yangtze above Nanking). From the observations we were able to make firsthand, however, it would seem that deep well pumping from groundwater has afforded a large portion of the water for irrigation in most of the provinces.

At every site that we visited, very comprehensive land shaping has been and is being done for effective irrigation water spreading on the soil. Much of the land is watered by a basin-flooding method with lesser application of deep furrow distribution. The continued land-leveling operation for countrywide irrigation is an important part of the capital construction program for agricultural development. It eliminates the traditional slack season of wintertime for the peasants, with everyone now involved during those months in moving soil by heavy machine, cart, and hand-carried basket.

A small amount of solid set sprinkler irrigation was observed as well, generally in association with the brigade experiment stations. Especially noteworthy is the amount of underground supply pipe that has been installed with risers at strategic points for water release. For

FIGURE 1 Basin flood irrigation of wheat, Shuang Chiao Commune near Peking.

example, the Shuang Chiao (Double Bridge) Commune near Peking, the Ma Lu Commune near Shanghai, and the Tung Yang Shih Brigade near Shihchiachuang appeared to have most of their water distribution systems so organized.

With the number of irrigation wells observed and the expressed installation depths of 60-100 m over much of the North China Plain, it is evident that a very large amount of electrical energy has been committed to the irrigation enterprise. The low relative cost of electrical energy for pumping of $10/ha, as quoted to us, has certainly encouraged the extension of deep-well irrigation by the communes.

The depth of the wells suggests that the aquifer is either very deep or very thick under the Peking Plain (by all appearances, a coastal plain, alternatively, an ancient floodplain, of the Yellow River). If the wells are necessarily deep to reach the water table, it seems probable that the aquifer is of bedrock, in which case a question arises as to the magnitude of the water level decline with the extensive pumping

FIGURE 2 High-yield plot of wheat (foreground) and sprinkler irrigation of wheat in background, Shuang Chiao Commune near Peking.

FIGURE 3 Pumping irrigation water from the groundwater and a return flow ditch into an underground distribution system, Shuang Chiao Commune near Peking.

now taking place for the multiple-cropping systems employed.

We were told that the amount and frequency of irrigation for wheat was determined by the crop water requirement and the environmental conditions. A rather set pattern was expressed, however, of a specific number and rate of irrigations in each province. More often than not, the water applied in an individual irrigation was specified as 30-35 m^3/mu (approximately 1/2 ac-in.; see Appendix K, conversion table). With eight (Kirin Province) to 12 (Shensi) irrigations during the crop's growing season the watering would be characterized as very light and frequent and highly labor intensive by U.S. standards.

In the brigade-level laboratories we were shown tables on lift and flow of irrigation water, suggesting comprehensive acquaintance of the responsible technicians with the engineering phases of water distribution. We did not at any time, however, have contact with an individual who might be classified as a soil water physicist. Nor did we encounter any specific data on evapotranspiration with different crops under varied environmental conditions as a base for crop irrigation. Neither could we obtain an estimate of water consumption with intercropping compared with a solid stand of a single crop. Rather, on inquiring about the manner in which the irrigation regime was determined, we were informed that watering was based on the experience of the peasant farmer. This has left us with considerable reservations concerning water use efficiency by crops. Excessive percolation appeared likely on the sandier soils at some sites where the routine was followed, while in soils of high clay content the light, frequent rate employed could be responsible for excessive evaporation loss. In the deep, loess-derived soils of Shensi Province, for example, we can visualize no logical basis for 12 or 13 irrigations of the wheat crop. No allowance is given thereby for the considerable moisture storage capacity of these deep, medium-textured soils. Furthermore, the crop is encouraged to root only to shallow depth, the amount of subsoil nutrient feeding that might otherwise occur with deeper rooting being reduced.

Serious thought is being given to the question of groundwater depletion by the responsible authorities at this time. Although we were never addressed by qualified hydrologists during the study tour, our hosts in Peking and in Kirin Province advised that unified plans have been developed for the spacing of irrigation wells. The matter is of some concern in the Peking area, where groundwater levels have indeed dropped noticeably in 1976 (10 m mentioned at two locations). The cause was expressed as the consequence of the limited rainfall since August 1975, but considering the average depth of wells, an overdevelopment with excessive withdrawal from the irrigation practice seems probable.

A very large potential apparently remains for groundwater exploitation on the high plains of Kirin Province. We were advised that the aquifer is deep, requiring wells of around 100-m depth, but contains a large supply of high-quality water. As a sufficient energy source becomes available and the wells can be dug and the land leveled, by reason of the high-quality soil in the region, bountiful crop production there will be assured for the future.

Irrigation rates described notwithstanding, many fields were observed from the train between Sian and Peking with existing irrigation ditches that displayed moisture stress. In many cases the moisture deficiency, compounded with nutrient shortages, was of such magnitude that very poor yields could be expected at best. Our hosts in Peking acknowledged serious limitations in irrigation development in the western portion of Hopei Province and expressed the governmental intent of correcting the situation as soon as possible. Noting the job that has already been done in harnessing the Hai River for flood control and irrigation and the magnitude of well irrigation already achieved on the Peking Plain, we have no cause to doubt the eventual realization of this objective.

Except for the more humid southeastern portion of the country, irrigation is regarded as the primary source of water in producing the wheat crop. In the shorter growing season of the northeastern region, specifically Kirin Province, and presumably in other similar climatic situations, irrigation practice assumes a supplementary role. Thus irrigation of the wheat crop is effected but not of the intercropped corn or soybeans, which receive adequate moisture during the high rainfall months of July and August.

Dry Farming

It was not our privilege to look in on any of the considerable dry farming areas that must exist in China. Only from the train and plane windows were we exposed to a small segment of this phase of the country's agriculture. A comprehensive study of wheat production in the non-irrigated portions of Szechwan, Tibet, Kansu, Shensi, Shansi, Hopei, eastern Inner Mongolia, Liaoning, Kirin, and Heilungkiang provinces would have been invaluable for the total perspective of this wheat studies team, especially in consideration of the fact that most American wheats are produced under rain-fed conditions.

Only in an early briefing session on wheat research and production and a final one after our tour through the country was the subject of dry farming touched upon. The experience of Tachai in moisture

FIGURE 4 Nonproductive irrigated wheat observed from the train between Sian and Peking.

conservation was elaborated with emphasis on the role of level bench terraces and deep plowing for accomplishing the objective in corn production. But judging by our fleeting observations from train windows, there remains much to be learned for dryland wheat production. The issue was passed over lightly by our Peking hosts, but there was some rather intensive questioning about U.S. experience with tillage methods in dry farming and summer fallowing in at least two of the institutes and academies visited.

Shelterbelts

A tremendous amount of tree planting has been done in the country since Liberation. A great deal of this has been as single-row shelterbelts around field boundaries. In the more southerly provinces these appear to be intended especially for fuel production purposes. But in the northern regions where a larger proportion of the land is nonirrigated, we were advised that the plantings were intended to serve a moisture conservation role by reducing wind action at the soil surface and thereby curbing evaporation loss of water.

It is common to observe wind barriers of corn and sorghum stalks (locally, wood slats or plastic) surrounding small fields adjacent to the cities in the northern portion of the country. These barriers are constructed especially for the protection of vegetable crops against wind damage but serve the added role of moisture conservation as well.

SOIL FERTILITY MANAGEMENT

China has long championed the art of organic manuring of soils and serves as a very good example for the rest of the world in this respect. Unquestionably, the practice has been fundamentally responsible for the generally high level of soil productivity that has been maintained through centuries of cropping. Even today with relatively limited use of inorganic fertilizers compared with consumption in Europe or North America, little visual evidence exists of nutritional deficiency in the crops observed on this study tour. It is recognized, of course, that our view has been somewhat distorted and may not represent the whole of China, since most of our exposure was to wheat production on alluvial soils. The latter are almost always the most productive among all types in any physiographic or ecological region.

Human and Animal Wastes

It is a revelation to a Westerner to observe the care with which all organic wastes are preserved in China. Nothing is wasted; if it is not eaten, the organic substance is returned to "strengthen the soil." The continuous traffic in barrels on carts and tank trucks carrying sewage from the city to the countryside, the gathering of horse droppings from the street, the night soil pit or pond by the rice paddy, the reverential attitude toward the hog as a fertilizer source--all evidence the national preoccupation with preserving this organic source of plant nutrients. As Mao has said "every pig is a fertilizer factory." Inquiry concerning measures employed for excluding the wastes of industrial plants in urban areas with heavy-element-toxicity potential afforded no answer. With the rush toward producing sufficient food for the masses this aspect of environmental quality and contamination of the food chain is apparently being disregarded.

Composting wastes is practiced everywhere. The stable manure, usually showing evidence of straw or stover admixture (accounting for some unknown percentage of crop residues removed from the soil at harvest), green manure, and night soil, is commonly mixed with a portion of stream or ditch sediment, if not field soil, and left to ferment for a period to accomplish nutrient mineralization (especially nitrogen). The admixture of organic and soil materials presumably provides the microorganisms required for the decomposition process, although inoculation of the compost heap with microorganisms and the "revitalization" of these organisms by irradiation was alluded to by the Institute of Atomic Energy Use in Plant Breeding (Peking). No specific organism type was indicated to us, and we were not convinced.

An unusual system of handling the organic compost was encountered in Kirin Province in the vicinity of Kungchuling. Here it is common practice to add black soil to the pig sty to serve as an absorbent of the urine. The soil plus pig droppings are removed every few days and replaced with fresh soil. On removal from the sty the mixture is piled, mixed with night soil, sealed with mud, and allowed to ferment. The major portion of a year is involved in the making of the compost, with application in the spring as the land is being prepared for cropping. Of the order of two thirds of the resulting heap is soil rather than straw or stover, the latter being used largely as fuel material in this cool climate if not as fencing material in the case of corn and sorghum stalks.

Inquiry into the requirements for organic compost formulation revealed a very substantial time commitment of around 8 man-days per mu in the middle provinces. This compares with variously quoted figures of 12-20 man-days per mu for the production of wheat. The large input for compost can perhaps be justified only on the basis of full utilization at all times of the labor that is available.

A moot question that always exists regarding use of raw human sewage for crop fertilization concerns the potential for spreading diseases. The issue was discussed by our hosts at their own volition in respect to the near eradication of schistosomiasis in the Ma Lu Commune near Shanghai. This "snail fever" disease was present in almost two thirds of the 14,000 members of the commune examined in 1958. A major feature of the control program involved the more comprehensive control of night soil, impounding it in only a few concentrated places. Subsequent fermentation in the controlled containing systems would kill the parasite responsible for transmitting the disease among humans. Educating the peasants to drink only well water and not that from streams or ditches further reduced intake of the organism to the extent that only 20 cases of schistosomiasis were diagnosed in the 29,000 residents of 1975.

FIGURE 5 Small two-wheel farm tractor pulling trailer with tank of sewage waste to composting site, Chiang Ning Commune near Nanking.

Many tropical countries of the world afflicted with this malady could benefit from China's experience with its control.

Green Manures

The leguminous green manures continue to serve a very significant role in supplying nitrogen for the cropping systems employed in the more humid and irrigated regions, especially in southern China with its longer growing season. An ancillary reason expressed to us was the importance of such fresh organic return for the amelioration of soil structure, probably realistic in consideration of the fact that much of the residue of grain crops is used for other purposes. With the shorter growing season of northern China there is much less dependence on green manure legumes. Kirin Province, for example, has less than 1 percent of its arable land in green manure crops and a noticeably greater willingness on the part of agricultural technicians to accept inorganic fertilizers as a major source of nitrogen.

In the middle and southern provinces, green manure crops are grown not only on the soil but in ponded basins as well. The latter crop is a fern (*Azolla*) capable of nitrogen fixation, probably complemented by blue-green algae, which is harvested from the water surface and subsequently incorporated into a field receiving its primary tillage for the crop to follow. The water hyacinth is harvested as well for this purpose (and for duck and pig feed) but presumably does not possess any nitrogen-fixing capability.

The Chinese were obviously interested in finding new sources of symbiotic nitrogen fixation. A project at the Institute of Botany of Academia Sinica in Peking has focused on nitrogen fixation on corn leaves by an *Azotobacter* sp. The only aspects of the study shared with us were the steps used to isolate, purify, and crystalize the nitrogenase enzyme from the *Azotobacter*. This work was apparently started only recently. At the Institute of Plant Physiology in Shanghai the acetylene reduction activity (a measure of capacity to fix nitrogen) of *Cycas revoluta* ThB. and *C. siamensis* Mg., whose coralloid roots are infected with blue-green algae, and several *Elaearnus* species (*E. pungens*,

E. multiflora, E. angustiflora, and *E. glabia*), having root nodules infected with an actinomycete, was being studied. These were mostly woody plants and would appear to have little direct application in most Chinese agricultural situations.

The actinomycete/*Elaearnus* species association was shown to reduce 0.19 μmol of acetylene per milligram of fresh weight per minute, while the alga/*Cycas* species had a lower activity. The project has also emphasized ultrastructure of the nodules by electron microscopy and has revealed that the actinomycete has branched, septate, intercellular hyphae and intracellular vesicles in the host. Other nitrogen-fixing symbionts marked for future study included *Podocarpus macrophyllus* and *Casnarina equistifolia*. All the plants under study in this project are native to China.

The green manure legume presents a contradiction to the "cropping index" or multiple-cropping objective now being stressed. Its growth period on the land prevents food crop production during that period (20-30 percent land occupation was variously expressed except for the far north provinces). Thus in the Nanking region, when milk vetch is seeded as the rice paddy is drained in the third crop of wheat-rice-rice, it will be possible to produce only two crops in the next year after its incorporation as green manure in the spring. Questioning of the responsible soil/crop management personnel at the various sites visited evidenced no consideration of the impact that the 3 plus million tons of fertilizer N from plants now under construction is likely to have on the green manuring practice. The existing attitude seems to be that the role of the green manure crop is inviolate, absolutely essential for maintaining soil productivity, not unlike the attitude of U.S. farmers prior to the 1950's. Neither does there seem to be appreciation of the high water requirement of legumes for those regions where crop production is accomplished under conditions of limited moisture supply, irrigation, or rain feeding.

The extreme importance of organic fertilizers in Chinese agriculture is unquestionable. The frequently expressed figure of some 70 percent of applied fertilizer nutrients coming from organic sources seems high until one considers the magnitude of a 10 ton/mu compost application for the wheat or rice crop. There can be no standard figures for the nutrient contents of compost, but even if the low values of 3, 2, and 3 kg/ton of available N, P_2O_5, and K_2O, respectively, were assumed, the per hectare treatment would be 450 + 300 + 450 kg of the three nutrients, to say nothing of all the secondary and micronutrients contained. Under the further assumption of a release of about one third of these nutrients in the first cropping season following application, the crop so treated is quite richly endowed with nutrients. Under such circumstances it is not surprising that nutrient deficiencies were rarely apparent in the fields visited during this study tour.

With such heavy dependence on organic wastes for maintaining soil productivity it is not likely that China will experience some of the undesirable changes in soil properties that have accompanied intensive agriculture in certain other soil regions. Deterioration in soil structure from heavy implement traffic is not likely to be so serious here with the continuous substantial return of organic materials. Buffered with silt, the compost will not have nearly as great an acidifying potential as would the same amount of nitrogen from an inorganic fertilizer source. Nor are such maladies as zinc and iron deficiencies induced by excessive phosphate fertilizer application likely to occur in China where the three elements are provided in reasonably balanced proportions in the organic fertilizer source. This is undoubtedly the reason why we saw so little evidence of trace element deficiencies other than in fruit crops at our many stopping points despite the generally calcareous nature of the soil. Organic farming is virtually a religion in this vast country and is deserving of our admiration.

Inorganic Fertilizers

The preoccupation with organic fertilizers and the lessons from Tachai have been responsible for keeping inorganic fertilizer in a subordinate position to this time. (No doubt the limitation in foreign exchange for effecting external purchase has been the underlying cause despite China's role as the major outlet for inorganic fertilizer in international commerce since the late 1960's.) Many of the agricultural production team and brigade leaders expressed to us suspicion of the use of inorganic N, based on the belief that such materials will cause a hardening of soils; perhaps there were other reasons not expressed. In view of the known dispersing action that the ammonium ion can have on soil colloids it is quite possible that deflocculation and hardening of soil, with nonjudicious use of some ammonium-base materials, has been observed. We are inclined to believe, however, that the distrust expressed is quite analogous to that of U.S. farmers when they were first exposed to inorganic nitrogen products in the late 1940's.

The above notwithstanding, nitrogen fertilizers are beginning to assume a significant role in the production of wheat in the communes of the country. Values of from 15-20 percent of the total N applied coming from inorganic sources were commonly expressed to us. Actual rates indicated ranged from 20-90 jin/mu of NH_4HCO_3 (approximately 20-90 lb of N per acre).

Inorganic fertilizer N is generally applied for wheat in the spring after organic dressings at planting and in the winter (mulch for plant protection against cold). Wherever any significant amount of the inorganic source is being used, as in excess of 5 jin of N per mu, split treatments are made with half in early March and the remainder in April, thereby providing the N just before the period of peak demand and minimizing potential leaching losses.

Among chemical carriers of N employed, NH_4HCO_3 appeared to predominate. It is manufactured in plants at the county or commune level by combining CO_2 with NH_3 received from state factories. The technicians who elaborated fertilizer use practices to us usually expressed the need for incorporation of this material into the soil for minimizing NH_3 volatilization losses, which could be expected to be serious from surface application to the generally calcareous soils of China.

Ammonium nitrate as an important fertilizer nitrogen source was encountered only in Kirin Province. A plant for its manufacture exists in Harbin. The local specialists had obviously accepted its use and the role of inorganic nitrogen more willingly than soil/crop management specialists in the other regions visited. Some communes seemed to have ammonium sulfate as the standard fertilizer nitrogen carrier. Presumably, this is largely material obtained by the Chinese government from foreign sources.

We encountered only one production group using urea fertilizer, although evidence exists of its purchase by China from international suppliers. Nor did we observe any serious experimental work with urea as the nitrogen source for wheat despite the large quantities of this material that will appear with completion of the 10 large nitrogen plants now being built. Rather, a suspicious attitude was noted among some of the soil fertility specialists concerning this "extremely high analysis" product. A preference was expressed for the bicarbonate product over urea at the Tung Yang Shih Brigade, despite a lower cost per unit of nitrogen in urea, because the ammonium bicarbonate is produced in a local county plant. Little question exists in our minds concerning the country's ability to absorb the large amount of urea in the offing, however, once it is readily accessible to the farmers.

Ammonia solution (aqua ammonia) is being used in some of the provinces. We observed a number of rather large concrete cylinders buried below ground level in the middle provinces that contain "liquid ammonia" dispensed by tube into irrigation furrows according to information supplied to us. This material was not being used at any of the sites visited in Kiangsu, Shensi, and Hopei provinces, however, and we have no idea of its relative importance in those areas. But in Kirin Province it is obvious that the ammonia solution is a very important source of fertilizer N because of the numerous large earthen pots in the fields for its storage. The material placed in the pots prior to the growing season contains 17-18 percent N requiring that the containers be sealed with plastic covers. When the container is opened for use, the solution is first diluted with additional water to minimize volatilization, then mixed with the organic fertilizer plus superphosphate programed for the field, followed by rapid incorporation of the complete mixture into the soil.

Although the traditional use of organic fertilizers has been invaluable for the maintenance of soil phosphorus levels in the soils of the country, the recently introduced technology for higher wheat yields has also exposed the general limitations of this element. The calcareous nature of most of the irrigated soils accentuates the need. As a consequence we encountered very few locations where phosphate fertilizer was not a part of the total production prescription. The technicians encountered confirmed the greater need of wheat for supplemental P than that of paddy rice on any specified field.

China apparently possesses significant rock phosphate deposits in Sinkiang and Yunnan provinces such that the raw material is available to the communes. They in turn often have their own chemical fertilizer plants capable of acidulating the rock with sulfuric acid to produce superphosphate. A relatively impure form of single superphosphate is the end product containing 16-18 percent P_2O_5. It appears to be rather common practice to apply 50-100 jin/mu of this material near or with the seed of wheat at planting.

An average expressed response of wheat to applied nitrogen and phosphorus fertilizers in the provinces visited was 3 1/2 jin of grain per jin of fertilizer as ammonium bicarbonate and superphosphate, higher in Kirin Province, where the dependence on organic sources of nutrients is not so great. With nutrient contents of approximately 17 percent N and 17 percent P_2O_5 in the above carriers the response per unit of nutrient is around 21, which is certainly highly acceptable by Western standards.

The most unique fertilizer formulation encountered during our study tour was at the Academy of Agricultural Sciences in Kirin Province, where ammonium nitrate, superphosphate, and the local compost are mixed in equal proportions. The resulting product is compressed by machine into granules of approximately 1/2-cm diameter. The contended enhancement of nitrogen and phosphorus availability from this granule is probably not without foundation. Proximity of the NH_4 ion to the phosphate could be beneficial to uptake of the P, and the organic presence might well extend the solubility and plant availability of the applied P.

Undoubtedly, potassium deficient soils exist in China, but none were encountered at any of our points of visit. The methodical return of organic wastes would seemingly afford partial explanation for the generally favorable status of this element in the soils observed.

There was little visual evidence of secondary or micronutrient deficiencies in any of the fields visited or in those observed from the car or train in our travels. We were advised that Mn is needed on some calcareous soils of northern Kiangsu Province, with correction effected by $MnSO_4$ at the rate of 2 kg/mu in solution form, alternatively by treating seed so as to supply 50 g of the salt per mu. The need for Mo was also expressed for certain soils of Hopei Province, with correction afforded by seed treatment with a sodium molybdate solution. The limited extent of shortages other than in N and P of the long-cropped soils of China, especially in view of their usually calcareous nature, is explainable only in terms of the continuous replenishment of the various nutrients by the organic manuring practice.

The most surprising feature of the inorganic fertilizer routine to us was the practice of foliar application of small quantities of nutrients in solution form to wheat at the flowering stage. Treatments varied among communes, but primary emphasis was on P. At several locations, KH_2PO_4 is the nutrient carrier, made up in a dilute solution of around 900:1 and sprayed at a rate to supply approximately 1/5 jin of the carrier per mu (an acre rate of around 0.3 lb K_2O and 0.5 lb P_2O_5). Other brigades were using superphosphate as the solute at a somewhat comparable rate of P_2O_5, and some were complementing this with urea at rates of 1/2 to 1 kg/mu. At one production brigade the K and P were obtained as a leachate from the composting of animal manure and wood ashes. In all cases the foliar treatment was claimed to keep the leaves functional longer, resulting in greater thousand kernel weights of grain. Yield increases from 3-15 percent were claimed at the approximate 300-350 kg/mu production level.

Evaluation of Soil Fertility Status

Our inquiries concerning the manner in which kind and amount of nutrients required on the soils of different communes were determined commonly received the answer that the peasant farmer knows the soils and their requirements very well. In essence, experience has taught him what is needed. Frequently, however, we were shown small soil testing laboratories equipped with quick testing kits of the type prepared by the Institute of Soil Science at Nanking. These involved reagents for measuring pH and levels of N, P, and K by colorimetric means evaluated by eye. Extracting methods employed are similar to those used in many U.S. laboratories with the $NaHCO_3$ extraction of P being standard over most of the country. Educated (intellectual) youths were usually indicated as being responsible for doing the analytical work, and we were informed that results from analyses of surface soil samples in these laboratories were used routinely for determining nutrient needs of fields in the brigade or production team.

In at least some of the provinces, county laboratories are responsible for testing the soil in most fields every 2 or 3 years. We were never afforded the opportunity of inspecting such a laboratory for evaluation of the quality of its instrumentation or the qualifications of its technicians. It probably goes without saying that we would not place much faith in the results from the laboratories observed, not because of lack of enthusiasm or integrity of those doing the work but rather because of the difficulties of effecting quantitative chemical measurements without benefit of electronic instruments or regular standardization of reagents. The intricacies of effecting correlation and calibration of the tests themselves to the nutritional status of crops in the field from one region to another impose further doubts of credibility.

Our hosts indicated that fertilizer trials were made routinely on representative soils of each region. This would, of course, give some indication of likely nutrient needs of similar soils with the same management history. It may well be that many such trials have been run in the recent past, but in all of our visits we encountered only one experiment involving time and rate of application of nitrogen, phosphorus, and potassium fertilizers, that in Kirin Province.

From all that has been seen it is evident that a better job of nutrient prescription than that presently being accomplished could be done in many areas of China. Some fields are receiving more than enough nutrients, and many others (observed from train and car) not enough. As inorganic fertilizers become more readily available for use on the fields in the next few years, a more comprehensive nutrient evaluation of soils will give multiplied returns over costs for the effort. Not only will economic interests be assured by providing additional nutrients where needed and eliminating excesses elsewhere, but the interests of environmental quality will be served as well.

We encountered no evidence that investigations are being made on the nutrient contamination of groundwaters and surface waters resulting from fertilizer practices. Frequent inquiry on the issue afforded answers of either "no information" or "this is no problem because most of the fertilizer used in China is organic." Visual evidence of surface water eutrophication exists everywhere, but much of this is intentional for producing aquatic green manure as described in an earlier section. But there is undoubtedly substantial nitrate movement into shallow groundwaters in valley positions wherever the very heavy applications of organic fertilizers previously elaborated are employed. In any case, the topic engendered considerable individual inquiry and discussion of U.S. experience on the matter with the Chinese scientists accompanying our delegation.

TILLAGE METHODS

Primary tillage is nearly everywhere accomplished by moldboard plowing in the regions visited. In some communes we were informed that the field is plowed twice in preparing the land for wheat, the purpose for which is not clear unless the second plowing is essentially a discing operation for firming up the cloddy surface from the first.

Plowing is considered absolutely essential, not only for loosening the seedling zone but for incorporating the copious quantities of organic manures employed. The agricultural production teams and brigade leaders always emphasized the need for deep plowing (15 cm) as one of the lessons from Tachai. Concession was obtained in Kirin Province that deep plowing could be restricted to alternate years without yield loss as an energy-conserving measure.

Secondary tillage varies considerably among regions with some row cultivation by tractor- or beast-drawn implement practiced where wheat is grown in widely spaced rows. Most commonly, however, the wheat is broadcast or seeded in approximate 15-cm rows with weed control effected by hoeing or hand rogueing.

CROP PRODUCTION SYSTEMS

Wheat is exceeded only by rice among crops produced in China, occupying about 20 percent of the arable land, and its importance as a component of the national diet is growing. Virtually all of the cropping systems encountered include a planting of wheat, winter or spring as governed by the climatic factor. One of the important reasons for this wide distribution from the subtropical to the northerly temperate zones must be attributed to the striving for self-reliance in every commune throughout the country. In so doing Chinese society is bracing itself against all possible calamities, including war (their oft voiced concern).

Environmental Determinants

Length of growing season, winter temperatures, soil quality, and available moisture are the more important environmental factors determining where and what type of wheat will be produced in the country and at what time in the year. Actually, China has demonstrated the capability of growing the crop in certain regions long considered to be quite inhospitable for wheat, e.g., on plateaus of Tibet reaching as high as 4,000 m where frost can occur almost any time during the year. In the major production area between the Yellow and Yangtze rivers the wheat type normally possesses a weak winter and spring habit and is fall sown to conform with the moderate severity of the winter. On the other hand, spring wheats are grown exclusively in the northeast provinces and at high altitude elsewhere, since winter types fall sown are subject to winter killing. At the opposite extreme in the far south provinces, spring types are sown in the summer, fall, or winter as best fits the cropping system employed.

An added reason for the continued expansion in wheat acreage in some communes is the growth in capability for irrigation of the crop. Much of China is in a twilight zone of rainfall for intensive agriculture with a major portion of that received coming in July and August. The dry weather in late spring is fine for the maturation and harvest of wheat, but the limited moisture conditions of early spring are quite unsatisfactory for the period of elongation and primordial initiation. Thus drouth effects are commonly experienced with both fall- and spring-seeded types (still apparent in dryland areas alluded to elsewhere in this report) until the availability of irrigation water for spring treatments alleviates the problem. Where little or no irrigation development has been possible to date in the drier regions, such as Inner Mongolia, there is very little wheat production, and the land is used rather for livestock-grazing purposes.

A further example of the impact of soil and climatic factors is the recent, developing role of a derivative of wheat in China, viz., triticale. This crop is taking the place of wheat and rye in mountainous regions of shallow, sandy soils due to its higher tolerance of moisture and nutrient shortages than that of wheat and its ready acceptance by the people as a food grain.

The length of the growing season does not prevent the growth of wheat in any but the higher altitudes in the western part of the country. This climatic feature, however, does govern the extent and kind of multiple cropping that can be practiced in any given location. With less than a 120-day season there is no real multiple cropping and only different versions of catch cropping or intercropping. Where the length of the growing season is not significantly reduced by frost, as in the area of Canton, three crops per year are now commonplace.

Intercropping

Several versions of intercropping, with wheat as one of the crops, are practiced in the various provinces, the type of intercropping being largely determined by climate and the needs for self-sufficiency in the particular locale. Catch cropping is a form of intercropping where a summer crop such as corn or cotton is planted into an existing stand of wheat that is approaching harvest. The second crop may be seeded or transplanted depending on the need established by the remaining length of the growing season.

Intercropping takes advantage of the border effects derived at the edges of fields where competition of contiguous plants for light, water, and fertility is reduced. Narrow strips of the two crops accomplish this objective, providing there are no pests that move into the later crop borders as the earlier crop is harvested and there are no hot summer winds to desiccate the border rows. Apparently, there are no serious maladies of these kinds to be contended with in China. The summer rainfall would appear to be the saving grace when results in China are compared with results for narrow strip cropping in many other temperate region countries. The summer crop in such a system will require little or no irrigation in a 500- to 600-mm rainfall region, as on much of the North China Plain and the northeast region. For example, in the Kungchuling area, grain production from a planting of alternate strips of approximately 1-m width of wheat and corn (two rows of 90,000 plants/ha), when compared with a solid stand of corn planted in a lower concentration of plants per row, produced the same amount of corn and, at the same time, yielded a 30 percent bonus of wheat. Similar results were reported for comparable interplanting of cotton with wheat in the vicinity of Shihchiachuang, the wheat there yielding about 50 percent of a full stand while allowing a maximum cotton yield.

Examples observed in the various provinces included wheat-corn, wheat-cotton, wheat-soybeans, wheat-sorghum, wheat-vegetables, and different combinations of the summer crops. One of the more striking combinations of the latter observed in many production fields of Kirin Province was the planting of soybeans in normal 7- to 75-cm rows, with corn planted in rows of around 150-cm spacing perpendicular to the soybean but in the soybean row to permit mechanical cultivation for weed control. As with other combinations an increase of about 30 percent in total production was claimed for the practice.

FIGURE 6 Wheat-corn intercropping research plots.

FIGURE 7 Wheat-cotton intercropping.

FIGURE 8 Barley-wheat-cotton intercropping.

Multiple Cropping

Wheat is being multiple cropped in almost every conceivable way in China. A few examples follow:

1. Wheat followed by cotton. The wheat is planted in beds approximately 3 ft wide in 15- to 20-cm rows. There is then a gap of 1 ft, in which the cotton is transplanted about 2 to 3 weeks before the wheat is harvested. We were told that wheat would later be seeded between the cotton rows about 3 weeks before the cotton was harvested. In that way, wheat that requires 210 days to mature and cotton that requires 200 days could both be grown on the same land in the same year. This system was widespread in the Shanghai area.

2. Wheat intercropped with barley. There were several versions of this combination in use around Nanking and Shanghai. The barley matures

about 2 weeks earlier than does the wheat. It is less valuable to the Chinese, however. Therefore, they plant two or three rows of barley between a few rows of wheat. When the barley is harvested, they transplant cotton. This lets the cotton become established while the wheat is ripening. After the wheat is harvested, the cotton occupies the entire field.

3. Wheat intercropped with maize. There were many versions of this combination in use throughout all of the wheat-growing regions that we visited. One version is to plant two rows of maize about 75 cm apart along the small irrigation ditches. This is done in early May in much of central China. The maize was from 10 to 20 cm tall at the time of our visit. The beds of wheat alternating with the double rows of maize were quite wide (3 m or more), and another crop would be seeded in the beds after the wheat was harvested (soybeans, rice, groundnuts, vegetables, etc., depending on the location). In the Shanghai area we saw many fields where this combination was used; the rows of maize were 120-day corn, and 80-day corn would be seeded into the wheat beds after wheat harvest. Another very common version of this combination was about 1-m wheat beds with a 30-cm gap in which 120-day corn was seeded about 3 weeks before the wheat was harvested. The field would then become a field of a single variety of maize.

4. Wheat intercropped with fruit trees. Almost any of the above combinations are also used in orchards. We saw many orchards (apple, pear, plum, peach, etc.) with different intercropping schemes utilizing wheat or with wheat alone. The wheat would be followed by another suitable crop in the latter case.

We were left with the impression that the "cropping index" or multiple cropping has become the number one consideration in plant breeding and all aspects of crop management. Early maturity is a must in all varietal development to accommodate as many crops per year on each mu of land the length of growing season will allow. The concept has been extended to the point of transplanting such crops as corn, soybeans, cotton, and wheat from the nursery beds, as has been traditionally practiced with rice.

The resulting combinations are numerous and all directed toward a sufficiency in food for the country as a whole and self-sufficiency for each administrative unit. In the provinces with growing seasons in excess of 200 days, as in those from Shanghai and Nanking southward, it is now fashionable to plant a crop of wheat in the fall with spring harvesting followed by two crops of rice in contrast with the traditional two crops of rice per year. The irrigation demand of wheat in much of this area of 100- to 200-cm annual rainfall is slight; rather, the number one water management consideration is that of artificial drainage for the wheat followed by flooding to create a paddy environment for the subsequent rice crop. Total grain yield from this combination in the communes visited was variously reported to be in the range of 8,000-10,000 kg/ha with wheat contributing around 3,000 of this total.

Northward from Peking with 160 days or less of growing season the conversion being made is from a single crop to two crops per year, with two and one-third crops obtainable as the combination of intercropping is introduced. The multiple feature is possible in these northerly provinces only as irrigation is practiced, of course, since the evapotranspiration requirement for any given crop here will range from 400 to 600 mm, which is the approximate mean annual rainfall. The planting of millet or soybeans after wheat was described to us in central Kirin Province, where the growing season is no more than 120 days, the practice requiring very early maturing varieties of the two crops. In other sites of limited season length there is interplanting of winter barley with wheat, allowing somewhat earlier transplanting of cotton in the barley stubble, while corn or sorghum follows the wheat.

Not only new varieties but new methods of cultivation are required as multiple cropping is extended to new areas. Some of those observed or described to us demonstrate the adaptability of Chinese farmers to change from traditional methods. The transplanting of corn and sorghum at the six-leaf stage (with success achieved even at the 10-leaf stage) permits maturing of the grain in only 60 days on much of the North China Plain. In places, wheat transplanting is made into the wet soil following the harvest of rice, after which the water is drained off. This permits wheat establishment very late in the growing season in areas where the season is barely long enough for maturation of the rice. Interestingly, transplanting of wheat is practiced in some communes for the express reason that yields are higher with transplanting than with conventional seeding (40 percent higher claimed in the Huai Ti brigade of Hopei Province), a result attributed to greater tillering with more heads produced. The routine involved liberal organic fertilizer treatment and irrigation of the nursery, transplanting after an approximate 45-day seedling stage with 1 mu of nursery supplying 4 mu of production field, setting the plants in groups of three per hill with hill-spacing of about 7 cm in 20-cm rows, and irrigation immediately after transplanting. The labor requirement for this operation is obviously tremendous but little different from that for transplanted rice and perhaps rational enough for this country of some 850 million population at its present stage of development. The same can be said for the transplanting of corn, sorghum, cotton, and soybeans, substantiating further the Westerner's impression of a garden type of farming practiced in much of China.

Our hosts expressed the intent of expanding the cropping index as rapidly as technology will allow, most particularly through the further development of irrigation facilities, the breeding of earlier maturing varieties of crops, and an increased production of fertilizer materials, especially nitrogen. A countrywide estimate of a cropping index somewhat over 1.5 at the present time with the goal of reaching 2.0 by the end of the decade is probably realistic if the government continues to place top priority on agriculture and, especially, irrigation development.

Residue Management

Although not a crop management practice per se, the manner of crop residue management has much to do with controlling erosion and the long-term preservation of high levels of crop productivity. The situation in China differs from more mechanized countries in temperate regions in the fact that most straw and stover are removed from the land in the harvest operation. In the central and southern provinces a major portion of these residues becomes part of the composted organic fertilizer returned to the land. In the north, however, much of the crop residue is used for fuel and, in the case of corn and sorghum stalks, for building yard fences. As mechanization of agriculture progresses in the country, the return of these fresh residues will become increasingly important for ameliorating the soil compaction brought about by heavy implement traffic. The animal manure and night soil combination probably will not provide the lignins needed for effective soil structure stabilization. Moreover, with expanding industrialization of the country and employment of an increasing proportion of the people in more enlightened positions, the very large labor input in compost production will be difficult to accommodate.

We can readily visualize a more effective use of residues than is now being made in dry farming wheat production of the country. Their full use in a stubble mulch tillage system would achieve the high soil permeability now accomplished by deep plowing while at the same time saving some moisture against evaporation loss and conserving substantial energy.

MECHANIZATION

Chinese agriculture is quite obviously in a transition stage as respects mechanization. Every commune visited indicated ownership of a few large tractors, several small walking tractors, and a few trucks. A contention of 80-90 percent of the land in the commune now being plowed with tractors was frequently expressed. We observed such machines in operation in all provinces as well as rather significant numbers of Caterpillar tractors doing the heavier plowing and land grading operations.

Plows ranged from the six-bottom moldboard type drawn by the larger tractors to those with a single share drawn by oxen, horses, or mules. Planting and cultivation equipment was of a similarly wide range, accommodating from several rows to single rows in the case of summer crops. It was obvious that a fair amount of the wheat was seeded by drills in row-spacing of 15-20 cm, although a substantial amount of it is still seeded by a hand broadcast operation.

Mechanical lifters of rice from the seedling nursery were observed in operation at many locations, as were mechanical planters of the seedlings. There were still large numbers of workers in the fields doing these operations by hand, however, and all of the transplanting of other crops observed was by hand.

A few field combines were observed, most of which were in the northern provinces, where fields tended to be larger. But most of the wheat harvesting is still carried out manually. Manual harvesting would appear to remain a necessity in most of those fields where the intercropping systems were employed. Threshing of the grain when so cut is achieved in many ways--by small beating machine, by animal trampling on a packed soil surface, by spreading on a hard surface road allowing

FIGURE 9 Grain-threshing floor with machines used for separating grain from heads at the Ma Lu Commune near Shanghai.

FIGURE 10 Barley threshing at the Ma Lu Commune.

FIGURE 11 Barley threshing by wheel traffic on a paved road near Sian. Note recent tree plantings.

highway traffic to separate grain from head--all requiring subsequent winnowing by hand.

In total, the human effort is still very great in carrying out the garden type of farming practiced in China. When everything is toted up, including the effort in compost preparation, around 25 man-days per mu are indicated as being expended in producing the wheat crop. Judging by the well-fed appearance of the entire populace, this intensive labor input made available by giving agriculture first priority in development could serve as an example for many developing countries presently experiencing food shortages.

WEED CONTROL

Weeds are normally the most damaging of all pests in the production of crops owing to their competition for nutrients, water, and light. Weeds did not appear to be a serious limitation to wheat yields at any of the sites that we visited in China, however. Although herbicides have been researched for Chinese conditions (everyone seemed to be aware of the potential of 2,4-D for weed control in wheat), we came across only a single instance in which 2,4-D was actually used for weed control in the production of wheat (a commune in suburban Peking). Thus, weed control in wheat in China can be summarized very simply--wheat fields are hand weeded.

We observed a number of crews weeding wheat fields. It appeared to be universal that the weeds were carefully collected (most workers carried a small basket to put the weeds in), and we were told that the weeds were taken home and fed to pigs. One suspects that in that sense, weeds are a "crop" and the use of herbicides would deprive the commune members of one source of feed for their pigs. Thus, although herbicides are available in China (one commune we visited had a rather large factory in which, we were told, herbicides were manufactured), they do not appear to be at all important in the culture of wheat.

At Shanghai we were told that the important weeds in wheat were: *Poa* spp., chickweed, field bindweed, foxtail, water foxtail, goosetail, and wild oats. In north China most weeds have been removed by hand. In northeast China, weeds were somewhat more plentiful but again were being adequately controlled in most fields by hand weeding. Wild oats were a major problem in the Sian area. The wild oat panicles were hand stripped in seed fields to remove the oats before harvest of the wheat. We saw this hand stripping in process in several fields. The worst weed problem observed was in northeast China, where recent wet and cool weather had not permitted any weeding. Most of those weeds were of the grassy type. The number of weeds in this area indicated a plentiful seed supply. Unless the weather permitted weeding in late June, losses in wheat yield could have been considerable.

At several of the institutes that we visited, there were divisions or laboratories evaluating herbicides. Apparently, most herbicides in use anywhere in the world are being synthesized in China. We encountered only one instance in which physiological reactions of herbicides on weeds were being studied. (The translocation of glycophosphate in nut sedge was being studied at the Institute of Plant Physiology in Shanghai.) Rather, the herbicide research was directed to practical screening for effectiveness and determination of rates to obtain good weed control.

PHOTOSYNTHESIS AND PLANT PRODUCTIVITY

Many of the wheat fields that we visited were highly productive. There was considerable interest among the Chinese research workers on whether yields were being limited by the processes of photosynthesis and photorespiration. The only place visited where photosynthesis of wheat genotypes was being measured, however, was at the Kiangsu Provincial Academy of Agricultural Research in Nanking. Our visit was rushed (total of one-half day), and we did not have the opportunity to see the laboratories, growth facilities, or equipment. We were told, however, that they had reasonably well-controlled environment chambers and that infrared analysis of carbon dioxide was being used to measure photosynthesis in both the field and the laboratory. The physiology group at the academy appeared to be working closely with the plant breeders, and we felt that they had a strong crop physiology program.

The Plant Physiology Institute in Shanghai was doing research on photosynthesis. They were measuring ATP synthesis rates. They were also doing some CO_2 enrichment in cold frames growing rice seedlings. The institute had a 25-room phytotron with xenon lighting, a very impressive facility for crop physiology research.

These two were the only places where work on photosynthesis was observed, although there was concern about leaf area and light absorption by wheat canopies in several places. We saw no micrometeorological equipment of any kind, except for a lux meter being designed by the instrument shop at the Institute of Genetics in Peking.

GROWTH REGULATORS

At the Institute of Botany in Peking, mention was made of tests of (2-chloroethyl) trimethyl ammonium chloride (CCC) on wheat. At the Sino-Albanian Friendship Commune we saw a high-yield field that had been sprayed with CCC. The expressed purpose was to decrease height and lodging; further, the sprays would delay maturity for a few days and increase kernel weights. Two applications of CCC had been applied at a rate of 10 g/mu at each application. We were told by the responsible person for crop culture that many communes in the Peking area used CCC on tall varieties. We were not able to verify that statement, and this one instance is the only example we found of the use of CCC. Specific questions about the use of CCC in other parts of China indicated that its use is not widespread.

We found no evidence of use of any other growth regulators on wheat or evidence of any research on this topic.

FIGURE 12 Greenhouse facilities with mobile tables at the Institute of Botany, Peking.

2

WHEAT BREEDING AND GENETICS

To fully appreciate the nature and scope of wheat breeding in the People's Republic of China, it is necessary to understand the relationship of research to the social, political, and cultural aspects of Chinese life. The desire to achieve self-sufficiency in all segments of the country including food production, the focus on class struggle, the aim of making research serve the immediate needs of the people, and the avoidance of establishing an elite class of scholars all have a direct bearing on how wheat breeding is currently being conducted. The overall effect has been to decentralize research with an almost exclusive thrust on those factors that offer promise of direct application to the existing problems. Research carried out in a so-called "open door" fashion or utilizing the three-in-one concept whereby the scientist, technician, and farmer all participate in a cooperative research program has resulted in the development and adoption of new technology in a very effective manner. In fact, this concept has resulted in the development of an outstanding extension system.

Wheat in China is second only to rice in importance as a food grain. Dramatic increases in wheat production have been realized in recent years as the result of an expansion and intensification of existing cultural practices, including a significant increase in the amount of irrigated land. Complementing these changes has been the development of superior wheat cultivars that represent standard height or semidwarf types with greater straw strength and that are more responsive to improved management practices. The first semidwarf cultivar was released in 1972 with the Chinese breeders utilizing the Suwon source of dwarfism originally obtained from Korea. Subsequently, a major effort to develop semidwarf wheats has resulted in the introduction of dwarfing genes from other sources.

Wheat is grown in all provinces in China covering an area from 18°N latitude near Hainan Island to approximately 50°N latitude. It is also found from below sea level to 4,100 m above sea level. A major portion of the annual rainfall is received during the summer months, the total amount varying greatly from one region to another. In most of the wheat-producing areas there is only limited snow cover during the winter months, and it does not remain on the ground for prolonged periods. In the northern provinces, dry, hot winds in the early fall limit the length of the growing season, but in other areas the duration is a function of

FIGURE 13 A highly productive irrigated field of semidwarf wheat.

the cropping sequence. As a consequence, early maturity is a prerequisite of all wheat cultivars grown in China. In addition, all the major diseases of wheat can be observed in China. Therefore, it is quite apparent that a major continuing effort in wheat breeding is required if cultivars are to be developed that can stabilize wheat production at a satisfactory yield level under such varying conditions.

Winter wheat represents approximately 85 percent of the total wheat production and is an important crop in the provinces of Kiangsu, Anhwei, Honan, Shantung, Hopei, Shansi, and Shensi and in the Shanghai municipality. The cultivars appear to be facultative types that do require some low temperatures to satisfy a slight vernalization requirement. There is no need for a high degree of winterhardiness, particularly near the southern limit of winter wheat production, which closely parallels the 30°N latitude. The northern limit of winter wheat production is established near the 40°N latitude; however, factors such as elevation and nearness to the ocean, as in Kiangsu Province, may modify the actual area where winter wheat can be grown. In these northern regions a fair degree of winterhardiness is required in addition to early maturity. Planting dates vary from September through November depending on the cropping practices, with harvest being completed during May and June.

The major production of spring-type wheats is found in the three northeastern provinces of Liaoning, Kirin, and Heilungkiang. This area extends northward from 41°N to 48°N latitude and has extremely cold winter temperatures similar to the wheat-growing regions observed in the Northern Great Plains in the United States and the provinces of Manitoba and Saskatchewan in Canada. Planting dates normally are April or early May with harvest in July or early August.

Spring wheats grown below the 30°N latitude are fall planted in November because of the mild winter temperatures. Due to double- and triple-cropping practices, especially with rice, early maturity of the wheat is a desired attribute, with harvest occurring in April and May. With the introduction of the day length-insensitive, semidwarf cultivars from Mexico, an expansion of wheat production in this area has been realized. The day length-insensitive characteristic has provided the

Chinese breeder with additional sources of earliness, making wheat more competitive with barley in the multiple-cropping sequence.

In both winter and spring cultivars there appears to be a preference for white grain; however, due to the frequent rainfall during harvest, many of the commercial cultivars have red grain. The logical explanation for cultivating red grain is to take advantage of the association of kernel color with postharvest dormancy, thus avoiding sprout damage. The high flour extraction percentage obtained during the milling process (85 percent or higher) no doubt contributes to the preference for white grain.

In most major producing areas there appeared to be two and sometimes three predominant cultivars in production. For winter wheat these cultivars can be characterized as semidwarf or standard in height with rather wide, nonerect, broad leaves and oblong-shaped spikes, including both awnless and awned types. The spring-type cultivars tended to be intermediate for most morphological characteristics.

One important attribute of the varieties currently in production is their apparent resistance to major diseases. Previously, China, like so many wheat-growing countries of the world, experienced frequent disease epidemics of stem rust (*Puccinia graminis tritici*) or stripe rust (*Puccinia striiformis*). However, the fact that no such problems have been encountered since 1964, suggests that the Chinese have successfully incorporated good sources of resistance into their material. Scab (*Fusarium* spp.) currently is the only major disease problem, and it appears that there are cultivar differences in degree of infection. Whether genetic factors for scab resistance are available or certain cultivars, owing to their date of flowering, are merely escaping the infection could not be ascertained. Certain cultivars have been tolerant over years, the suggestion being that some sources of genetic resistance may be present.

Wheat breeding, without question, has received a great deal of effort as the result of the emphasis on applied research. The new wheat cultivars and such factors as irrigation, more fertilizer, and different cropping systems, along with breakthroughs in other crops such as rice, have resulted in the Chinese feeding 850 million people while being restricted to the intensive cultivation of only 11 to 15 percent of their total land area. This is a major accomplishment, and much can be learned from the Chinese experience in increasing food production. The concern that must be raised is the possibility that such emphasis on applied research eventually will result in a void of basic information necessary for further progress. When this question was raised at various academies and institutes, it was suggested that this trend toward applied research was a temporary measure to solve the most pressing needs of the people and that a greater balance would be achieved with basic research at some future date. Since the educational system is also very much oriented toward applied work, one might question whether the delicate balance between population and food supply will ever permit such a reallocation of resources before new and more complex problems arise that require basic research information.

ORGANIZATION OF WHEAT-BREEDING RESEARCH

Generalizations are made regarding the various levels of responsibility for wheat-breeding and related research. This is done with the full realization that there may be numerous exceptions within and between provinces; however, a 4-week tour does not permit an adequate base to do more than generalize.

Wheat-breeding research is conducted in the Ministry of Agriculture, which encompasses the Chinese Academy of Agricultural and Forestry Sciences and various national laboratories and institutes. At the provincial level there are the academies of science for agriculture and the colleges of agriculture. These, in turn, are closely associated with various activities conducted at the county, commune, brigade, and production team levels. In addition, there is wheat-breeding and related research at the national level being conducted under the Academy of Science, which is separate from the Ministry of Agriculture and includes the Institutes of Botany, Genetics, and Microbiology. In all academies, institutes, and colleges, approximately one third of the staff are working at the commune, brigade, or production team levels at any point in time. The time that staff members spend working at these levels varies from a few weeks to several years, depending on their particular assignment. In addition to working in the actual production of food, they conduct seminars and short courses and work with the technicians and experienced farmers in planning and conducting wheat research. This interaction serves to (1) help the scientist understand the problems of the people, (2) assist in educating and training young people in breeding methods, and (3) extend scientific technology to the farmers.

The most comprehensive breeding programs visited were at the province level, involving various institutes and academies of agricultural sciences. Of those visited, Kiangsu Province Agricultural Research Institute in Nanking; the Shensi Province Academy of Agricultural and Forestry Sciences and the Northwest Agricultural College, both located at Wukung; the Hopei Cereal Crop Research Institute; the Kirin Academy of Agricultural Sciences; the Institute of Atomic Energy Laboratory in Peking; and the Shanghai Academy of Agricultural Science had interesting and noteworthy programs. It would have been desirable to have spent time in the experimental plots at the Institute of Genetics to observe the wheat and triticale experimental material; unfortunately, time did not permit such a visit. The scientists visited were very knowledgeable about their programs. It would appear that such academies and institutes are the main source of new cultivars and germ plasm provided for regional testing and they are responsible for sending segregating populations to the lower levels of research.

Wheat research at the county level appears to function mainly in a coordinating role for arranging meetings, disseminating material for regional yield trials, and establishing certification standards. Researchers at this level conduct field inspections to insure that varieties in commercial production remain genetically pure. We were told that in some instances, research activity is conducted at the county level; however, the extent and nature of such work was never made clear.

FIGURE 14 An attractive new semidwarf Chinese wheat variety at the Institute of Agricultural Sciences near Nanking.

Some wheat breeding is conducted at the commune level; however, for the most part, this effort appeared to be a modest one. The Shuang Chiao People's Commune near Peking had what appeared to be a small fairly productive program. This research effort was conducted in concert with scientists associated with the research institutes in Peking. However, most communes appeared to function more in an extension role such as coordinating meetings and research at the production brigade and production team levels. Communes together with the state farms also were involved in seed multiplication for some regions. At the brigade level, some breeding work was being done on a limited scale, with most of the activity involving some form of mass selection within existing cultivars. Where crosses were being made and plants selected from segregating populations, the objectives appeared to involve training rather than a productive breeding effort.

The production teams were responsible for maintaining the genetic purity of existing commercial cultivars. Head or plant rows were grown, and phenotypically similar types were bulked.

Yield trials, which were grown both at the brigade and production team levels usually, were comprised of nonreplicated observation plots. Experiment stations operating at the commune and brigade levels were conducting various agronomic trials on interplanting, seeding rates, and fertility in addition to some breeding work.

Recommendations regarding the release of new varieties, possible areas of adaptation, and seed multiplication result from meetings held at frequent intervals during the year. Depending on the province, these meetings are organized by the academies or institutes and, in some instances, at the county or commune level. Those participating in the breeding work at all levels were encouraged to participate in such meetings.

BREEDING OBJECTIVES

The observations made and conclusions drawn regarding the major objectives of wheat breeding in China are limited to the winter wheat programs visited in Peking and Shanghai municipalities, Kiangsu and Shensi provinces, and the spring wheat program in Kirin Province.

As has been previously noted, the one major objective of all programs visited is early maturity. In the winter wheat area of the Yangtze River Basin it would be desirable to have cultivars requiring 180 days or less from sowing to maturity, thus allowing the subsequent planting of two crops of rice within the same year. At Sian, the double cropping of wheat and corn or other crops would also be favored by earlier maturity of the wheat.

The short growing season in the northern spring wheat areas, which is due to the cold spring and the hot fall winds, also results in the need for early maturity. In the south, where mild winter temperatures permit the fall sowing of spring wheat, earliness gained through incorporating the genetic factors for day length insensitivity has also been utilized.

One of the major goals in China is to grow most of the wheat under irrigated conditions. Therefore, the development of semidwarf cultivars is receiving much attention. The desired height levels range from 80 to 90 cm. Suwon 88 from Korea apparently was one of the first sources of dwarfism utilized; however, now Tom Thumb, Norin 10, and Olsen sources are all being incorporated into the programs. Frequently, useful material is obtained from foreign cultivars in which short stature has been combined with other superior agronomic traits.

Some variation was noted in the different programs regarding the desired plant type. There did appear to be a trend toward blocky head types, with or without awns and with intermediate tillering capacity. Since rather high seeding rates are used in commercial production, high tillering capacity was not an important selection factor. Leaf width and position appeared to vary; however, it was suggested that where intercropping was practiced, there was some selection for upright leaves for greater photosynthetic efficiency. There is some question regarding this assumption, since it might be more important to have upright leaves in solid seedings. The Chinese winter wheat cultivars were similar to such varieties as Sava and Roussalka from eastern Europe and Hyslop from the Pacific Northwest, except they were earlier in maturity. The spring wheats were similar in appearance to many of the semidwarf types being developed by International Maize and Wheat Improvement Center (CIMMYT) in Mexico and in the Minnesota and North Dakota breeding programs in the United States; however, greater emphasis was placed on spike size in the Chinese germ plasm.

Breeding for disease resistance will continue to be a major objective in the breeding programs. The last major epidemic occurred in Shensi Province in 1964 and involved stripe rust. With nearly every major disease known to infect wheat being present in China and rather strong evidence that there have been some changes in the physiological race patterns, the breeders obviously have been effective in developing resistant cultivars. In the winter wheat region in Kiangsu Province and west to Sian in Shensi Province, the major disease is scab. It was reported that several sources of resistance had been identified in a number of cultivars including Frontana from Brazil, Vanissa from

Austria, Ming wheat no. 3, and Jin-Kwang from China. Jin-Kwang was used as a source of scab resistance for Selection 6701, an excellent agronomic type which also appeared to be resistant to leaf and stripe rust. Resistance to scab was not complete, and the material observed reflected a generalized type of resistance. The Chinese wheat breeders did indicate that these resistant lines were not escaping scab because of time of flowering but carried genetic resistance. The wheat-corn rotation, which is common in Shensi Province, certainly favors the development of the scab fungus. Rice also is a host for the scab fungus in the Shanghai and Nanking regions.

Stripe and leaf rust (*Puccinia recondita tritici*) are potentially serious disease problems; however, there appeared to be many sources of resistance in the programs. Powdery mildew (*Erysiphe graminis*), *Septoria tritici*, and *S. nodorum* could also present some problems along with *Helminthosporium* leaf blight; however, they were receiving no attention in the programs visited. Common bunt (*Tilletia caries* and *T. foetida*) has been brought under control with resistant varieties and chemicals. The environmental conditions did not appear favorable for the development of dwarf bunt (*T. controversa*). Flag smut (*Urocystis tritici*), although found, was not of major consequence.

A number of root- or crown-rotting fungi such as *Cercosporella herpotrichoides* were noted. It appears that the use of so much organic fertilizer and the related microbial activity may control these pathogens; however, if more inorganic fertilizer is applied, it may be necessary to breed for resistance at some future time.

Several virus diseases, including yellow dwarf virus, were being kept in check by spraying for aphid control. Only at the Kirin Academy of Agricultural Sciences was there concern for obtaining sources of resistance to yellow dwarf virus. Midges and armyworms along with aphids were present in some locations but apparently were not major problems.

METHODS OF BREEDING

Hybridization was conducted at all levels, ranging from institutes to production teams. However, the crossing blocks noted below the provincial level were small in size, and the amount of genetic diversity was very limited. Also, the number of crosses made each year appeared to be few in number with inadequate segregating population sizes for evaluation and selection. Wheat-breeding programs visited at the academy or institute level were more adequate not only in terms of the diversity of parental types but also in the number of crosses made each year. In addition to two-way crosses, considerable effort was being devoted to three-way and four-way crosses. With three-way crosses the third parent was usually a well-adapted Chinese cultivar.

There were also a number of winter × spring crosses being made with the specific objective of developing facultative or very early maturing wheats. In the program at the Kirin Academy of Agricultural Sciences, winter wheat was vernalized by first treating the seed with $HgCl_2$ for 30 min to sterilize the surface of the kernels. The seed then was washed and held for 30 days at 0°-3°C before transplanting into the field for crossing with spring type wheat.

Most foreign introductions were unadapted to the growing conditions in China, being late and susceptible to scab in the winter wheat areas or susceptible to cold injury in the northern spring areas. However, they were being used extensively in the crossing programs. This was particularly true for materials from Mexico, from which earliness could be achieved through day length insensitivity. Sources of stripe rust resistance were being obtained largely from European material. Leaf and stem rust resistance was being contributed by materials from the United States and Canada. It was interesting to note at Sian that many crosses had been made between local wheats and entries in the provincial germ plasm collection. However, apparently very little effort was being made to incorporate genetic material from the older land varieties of China.

The techniques of emasculation and pollination were similar to those commonly utilized in other countries. The florets were cut back, anthers removed, and a paper bag placed over the emasculated spike to prevent contamination from foreign pollen. The pollination was done either by the approach method or by cutting the top of the paper bag and twirling the pollen spike so that the pollen was shed onto the emasculated flowers. Where single crosses were involved, two to five heads were emasculated, while eight to ten spikes were used in three-way or four-way crosses. The number of cross combinations varied from a few to over 150 depending on the size of the program.

The method of handling segregating populations followed the pedigree system for the most part; however, some modifications were imposed in later generations. Usually, individual F_5 rows were bulked to provide material for yield testing in the F_6. However, at Kirin, a modified pedigree system was used whereby individual F_3 plant rows were bulked if the lines were fairly uniform. If not, individual plants were selected for the F_4 generation. Like most programs in the United States, the breeders had modified their methods of handling segregating populations to meet the objectives of their programs. For example, the breeders at the Kirin Academy of Agricultural Sciences backcrossed superior F_2 plants to the recurrent parent. They also followed a modified pedigree method in an effort to get large population sizes in the early generations for yield testing. It was also mentioned that in some breeding programs, segregating populations were released to other programs at the commune and brigade levels and, in some instances, to production teams.

Where three-way and four-way crosses were made, individual plant selections were made in the F_1. Generally, the population size in the F_2 was about 2,000 plants per cross; however, this varied with the specific program. In fact, at Shanghai, it was indicated that as many as 6,000 F_2 plants per cross were evaluated.

The segregating populations were subjected to various selection pressures. Spreader rows of lines susceptible to leaf and stripe rust were utilized; however, it did not appear that this material was always inoculated to insure infection. In the case of scab at Nanking, conidia spores were sprayed on the breeding material two to three times during the flowering period. Also, residue from infected barley was placed in the wheat plots 1 month before flowering.

Apparently, no screening for milling and baking quality or protein was conducted anywhere with early generation material. At some locations some quality tests were conducted on advanced lines, at least for protein. Apparently, the need for maximum production has replaced any consideration of quality as a selection criterion other than visual appearance of the kernels.

Owing to the great diversity of climates in China, the breeding programs may be accelerated by obtaining two and sometimes three generations per year. For example, spring wheat can be crossed in March near Peking and harvested at the end of June. The resulting F_1 seed can be planted in July in the mountainous areas of the southwest in Kweichow Province, where cool temperatures prevail through the summer months. In late October or the first of November, the F_2 seed can be harvested and planted in the Hainan Island area, with the F_3 generation seed being returned to Peking in time for spring planting. Most of the major spring wheat programs were utilizing other locations to advance the number of generations per year; however, two generations per year appeared to be the most common. It was not clear if any attempt had been made to accelerate the winter wheat programs.

Regional testing of F_5 and advanced lines was coordinated usually at the province level through the county and commune organizations and conducted by the brigade and production teams. Frequently, these were nonreplicated trials that also included foreign introductions. It was mentioned that replicated trials were conducted by experiment stations operated at the commune and brigade levels. The plot sizes varied for these trials, with 4 m in width and 20 m in length being common. There could be as many as 25 production teams, or more, within a given commune, all conducting some types of nonreplicated yield trials. When seed supplies were low, only a limited number of locations would be utilized for growing a new cultivar.

Each year at the province, county, or commune level, meetings are held to share information and make decisions regarding new cultivars and areas in which they are to be tested. As new cultivars are developed, experimental and demonstration trials are established, and farmers invited to the institutes to see the new material. Field days also are held in the countryside at sites of demonstration plantings. It would appear that the relationship between the scientist and the farmer is excellent and promotes early transfer of new information or adoption of new cultivars. In the Peking area a seed association handles seed increases. Once a decision has been made to release a new cultivar, the seed multiplication is organized and carried out by the state farms or at the county and commune levels.

Genetic purity of new or existing varieties is usually maintained at the production team level. Here, mass selection is practiced either on an individual head or plant row basis. The county establishes the standards of genetic purity and conducts the field inspections to insure that seed stocks continue to be representative of the cultivar.

Hybrid wheat was mentioned during discussions with Chinese scientists. However, there was very little indication that it was receiving much attention in China.

Considerable work was under way in which cobalt 60 was used as a source of radiation to induce mutations. Indications were that some success has been achieved in reducing plant height by this method. Apparently, no chemical mutagenic agents were used in mutation breeding. This was somewhat of a surprise, since the chemical mutagens usually are considered to be less drastic and more effective than is high-energy irradiation because they cause less chromosomal abnormalities.

ANTHER CULTURE

To shorten the time required to develop new cultivars, there was extensive use of anther culture. In fact, the anther culture technique was being employed at various levels of research, including the communes and brigades, by educated youths who had been taught the procedures by visiting scientists. Since this technique has not yet been fully developed, it is questionable how effective it will be as a practical tool at this time in Chinese wheat breeding. Unfortunately, there was no opportunity to see how the doubled homozygous progeny were being screened for various agronomic traits.

Pollen culture was observed at three institutes and one commune and was mentioned, but not observed, at several other institutions. Apparently, this method of obtaining immediate homozygosity of materials is considered to be of high importance. The method of obtaining doubled haploids varied slightly from one institute to another. Pollen culture apparently was started in 1970 or 1971 at the institutes. Rice has had a higher success rate than other food grains, but 10 species apparently have been successfully cultured. Wheat has a much lower success rate than rice.

The procedure was briefly described as follows: Murashige and Skoog medium was used initially for wheat, while Miller's medium with 2,4-D (1-2 ppm) and 9 percent sugar for callus formation was used for rice. After callus formation, the medium is changed to one containing a lower sugar percentage with lAA and a chelating agent added. After plant development, the material may be transferred to a third medium depending upon species and growth. Because NH_4^+ ions were found to adversely affect the rate of haploid induction, the concentration of NO_3^- ions was increased. This modified medium was called N6.

The success rate of haploid plants of wheat was claimed to be about 10 percent but varied widely and was genotype dependent. Haploid plants are treated with colchicine for chromosome doubling, and doubled tillers are selected. Several wheat cultivars (spring habit) were claimed to have been developed by this method. Scientists at the Institute of Genetics stated that 1,800 doubled wheat lines had been obtained in 1975.

Doubled haploids may speed the development of wheat cultivars using the Chinese method of yield testing. Pure lines could be rapidly distributed to the experimental stations and communes for yield testing without further selection for purity. However, the main disadvantages are that there is low frequency of success in wheat, no new genetic variability is created, and selection and testing still must be done.

SOMATIC HYBRIDIZATION

Somatic hybridization is employed by the Chinese as a possible method to enhance genetic variability. Although the probability of success is low, the potential of combining widely divergent species and genera could be of great importance. The first requirement for successful somatic hybridization is the technical ability to reproduce the entire organism from a somatic cell. Work at two institutes on somatic hybridization and tissue culture was presented, and some progress was indicated. Cell walls have been successfully removed from cells in the root, stem, and leaves of about 10 species (radish, beans, peas, tobacco, wheat, rice, etc.). Tobacco has been regenerated to a plant, but callus tissue has not been obtained from any other species. One cell division of wheat has been obtained using Murashige and Skoog media. Cell fusion was observed in photomicrographs, and in some cases, true fusion seemed to have occurred. In other photomicrographs, both nuclei were together, but the nuclear membranes were still intact. No cell divisions have been observed in the fused cells of species under study at this time.

TRITICALE

Chinese triticale research was discussed by Mr. Pao Wen-k'uei and Mr. Sun Yuan-shu of the Peking Academy of Agricultural Sciences. Octoploid-level triticale has been under development, in contrast to the hexaploid-level development in most other programs around the world. Many problems such as kernel quality, sterility, and tall plant types were noted that are similar to the experience of others working with triticale.

PROTOTYPE TRITICALE PRODUCTION

Initial steps to produce triticale require an abundant supply of genetic material. Since triticale is a man-produced species, no natural genetic variability exists in nature. Most common wheats cross poorly with rye. However, "Chinese Spring" wheat will cross with rye and produce rather high frequencies of seed set. Crossability with rye was being studied. Wheat cultivars are the determining factor in seed set, not the rye. Wheat was used exclusively as the female parent to avoid the rye cytoplasm.

Wheats were broadly classified into several categories for ease of crossing with rye. The categories were reported to be the following:

1. *Triticum sphaerococcum* types: low initial success as a direct cross (approximately 18 percent). The success of crossing was nearly doubled (34 percent) when *T. sphaerococcum* × Chinese Spring F_1 was crossed to rye, thus using Chinese Spring as a bridge.
2. "Ardito": a wheat from which direct crosses with rye produced a low rate of success (<3 percent). When Chinese Spring was used as a bridge, and the F_1 crossed to rye, successful cross frequency was

increased to approximately 40 percent.

3. "Nanda 2149": in direct crosses with rye, success frequencies were very low (<1 percent). When it was crossed with Chinese Spring as a bridge and the F_1 was crossed with rye, the success frequency was enhanced but was still relatively low (<14 percent).

4. "Quality": an example of a wheat almost impossible to hybridize with rye. No successful crosses were obtained directly with rye, and even (Chinese Spring × Quality F_1) × rye produced success frequency of less than 1 percent.

The wheats, if classified into ease of direct crossing with rye, would rank in this order: (1) Chinese Spring, (2) *T. sphaerococcum*, (3) Ardito, (4) Nanda 2149, and (5) Quality.

In order of dominance for crossability, which a later genetic analysis indicated was a multiple-allelic series, the varieties would rank in the following order: (1) Quality, (2) Nanda 2149, (3) *T. sphaerococcum*, (4) Ardito, and (5) Chinese Spring.

Using this genetic information, prototype triticales were produced by using Chinese Spring as a bridge. Rye was used as a tester for the F_1's and if seed set exceeded 10 percent, all spikes of the F_1's were kept for crossing with F_2 segregates with highest crossability. Usually, pure lines were not crossed with rye, but F_2's or later generations were used for genetic diversity and to ensure 20 percent hybrids as a minimum. After doubling, the hybrids were used in further crossing. By using this technique, over 5,000 prototype triticales were produced, about 2,000 of which, exhibiting good growth, were selected to provide genetic diversity.

The ABDR genome hybrids were doubled by using well-developed tillers during tillering growth stage, making a small cut in the stem of the tiller (not exceeding one half the diameter of the tiller) and applying a 0.05 percent solution of colchicine for 4 days, with the temperature held constant at 15°C. The tillers then were washed and transplanted to cold beds in which the temperature was maintained at 10°C with high humidity. A 90 percent survival rate of tillers was achieved by using this method, but no data were given on the frequency of successful doubling.

After successful doubling and selection of desired prototype triticales, selection pressure was applied for kernel plumpness and fertility. Early selections generally had low fertility and plump seed or higher fertility and shriveled seed, but none had both characteristics.

The main traits for selection were fertility and seed plumpness. Fertility was observed in the field with seed plumpness evaluated in the laboratory. A scale of seed characteristics was set from 1 to 5:

1. Unwrinkled seed coat, fully plump
2. Fully filled, seed coat wrinkled slightly
3. Well filled, seed coat more wrinkled
4. Seed shriveled in places and wrinkled
5. Ratio of bran to endosperm about 1:1

The lowest acceptable grade of seed was considered to be 3, similar to spring wheat in the Peking area.

Winter wheat in the Peking region was used as a fertility standard (87 percent fertile). Some triticales with 80 percent fertility have been identified.

Selection pressure for agronomic types was applied on the F_2 populations, which were seeded competitively to determine ability to tiller. Selected F_2's were seeded into F_3 head rows, with individual plants selected from the better rows. Yield testing of later generation material is conducted by the "peasant masses" in the marginal wheat production areas, where rye normally outyields wheat.

The desired types of triticale are tall, 1.3 to 1.9 m, normally spring type or facultative winter, and may be used for grain or forage. Winter habit triticales would not be grown extensively because of competition from winter wheat. If triticales are grown in the winter wheat area, forage would be the major use. The currently used Chinese triticales are later heading than are spring wheats. Some are resistant to powdery mildew, leaf rust, and stem rust.

Generally, triticale yield results have depended upon the environmental conditions in which they have been tested. Wheat normally outyields triticale in good environments, but wheat usually yields less than triticale or rye in medium and poor environments. Consequently, triticales may have promise as an alternative crop that may help to stabilize yields in the medium to poor environments.

The present triticales have a seed plumpness of 3; a few are type 2. They produce better steamed or baked bread than rye and have exhibited higher protein than winter wheat in the Peking area.

Some CIMMYT and University of Manitoba hexaploid triticales have been grown in China, but their seed plumpness was inferior to the Chinese triticales under the environments in which they were tested. Crosses with this material and the Chinese octoploid types could add another source of genetic diversity, especially for plant height.

GENETIC MATERIALS

Owing to the long and extensive cultivation of wheat in the many diverse climatic regions of China, much genetic variability has been created. Unfortunately, many of the very heterogeneous land varieties may have been lost and, with them, a wealth of important germ plasm. For example, the cultivar Chinese Spring, which contributed so much to the cytogenetic and genetic understanding of the wheat plant, was a selection made from a land variety of southwestern China, which has since been lost.

In 1955 a program was initiated to systematically collect, catalog, and preserve wheat cultivars in China. Germ plasm banks have been established at the province level. One collection of approximately 4,000 entries was being maintained at Wukung by the Shensi Province Academy of Agricultural and Forestry Sciences. Each year, about one fourth of this collection is grown to maintain seed viability. Included were materials representing some of the older cultivars grown in China. As is true for most long-established cultivars, these were tall with very weak straw. Lines appeared to be mainly hexaploids with a few

FIGURE 15 Propagation of Chinese wheat germ plasm at the Shensi Province Academy of Agricultural and Forestry Sciences near Wukung.

tetraploids. A few lines were observed that may have resulted from wheat/*Agropyron* crosses; however, it was unclear if these were naturally occurring or the result of a program directed toward wide crosses. More time for examination of this material and germ plasm collections maintained in other provinces would have been worthwhile.

A part of the collection consisted of more recently developed cultivars, with about one half being foreign introductions. The foreign material was mainly from Albania, Australia, Canada, Chile, England, France, Romania, Russia, Yugoslavia, and the United States. A high proportion of the material represented Eastern European type wheats. There were approximately 100 lines or cultivars developed in Mexico by the International Maize and Wheat Improvement Center. The collection consisted of both winter and spring types. However, since Shensi Province is mainly a winter wheat region, the collection was predominantly winter- or facultative-type wheats.

Sources of dwarfism, such as Tom Thumb and Norin 10, also were present in the material. It did not appear that the older Chinese cultivars were being evaluated for attributes such as new sources of disease resistance other than the noting of those that were resistant to the natural infections of scab and stripe rust. Crosses were being made to many foreign introductions that were mainly cultivars released 6 to 10 years ago. There was emphasis on short stiff straw and stripe rust resistance in the crosses.

One notable variety that had been introduced and used widely in Shensi Province on a commercial basis was called A-Po from Albania. This variety is probably too tall or weak strawed for highly fertilized irrigated production and was being replaced by shorter Chinese varieties, some developed from crosses with A-Po that were better adapted to those conditions. A-Po had a general appearance (somewhat like Glenlea, a Canadian spring wheat) of a tall Italian wheat with large awnless spikes, good stripe and fair leaf rust resistance, and high yield but with apparently poor grain quality.

The germ plasm collection maintained at the province level was available to all programs; however, most breeding programs were utilizing what seemed to be a rather limited number of parental lines in their crosses. This was especially true at the commune and brigade levels.

All germ plasm sent to the People's Republic of China from other countries must first be evaluated in Peking. Owing to concerns of introducing unwanted pests and diseases, all cultivars and segregating populations are carefully examined and evaluated prior to their dispersal to the various academies and institutes.

In the spring wheat region of Kirin Province at the Institute of Agricultural Sciences, Kungchuling, foreign introductions brought in by the Plant Studies Delegation were observed. The cultivars from the CIMMYT program were Penjamo 62, Pitic 62, Norteno 67, Tobari 66, Cajeme 71, Yecora 70, Azteca, Tanori 71, Saric, Noreste, Nadadores, and Nuri. Cultivars introduced from the United States and Canada were Chris, Polk, Era, Fletcher, Kitt, Waldron, Olaf, Ellar, Manitou, and Neepawa. Wheats from India and Pakistan were observed as well, most of which probably had some CIMMYT germ plasm in their pedigrees. These introduced materials should provide useful sources of disease resistance, short stature, and other desirable genes for further improvement of spring wheats for both irrigated and dryland production in China.

3

PESTS OF WHEAT

THE STATUS OF PLANT PATHOLOGY AND ENTOMOLOGY RELATED TO WHEAT IN CHINA

Plant pathology and entomology are combined into one discipline in China--plant protection. We agree with the statement by the Plant Studies Delegation that there are probably more plant protectionists in China than in any other country in the world. An estimate of the number is impossible, but the figure of 50,000 given earlier (Plant Studies Delegation report) is conservative.

As with other aspects of crop production and husbandry in China, plant protection, as applied to wheat, is largely an application of information known for at least the last 25-30 years. The main approach is through development of resistant varieties by conventional breeding. There is a particularly large effort to use pathogen-free seed and plant quarantine, administered at the county and provincial levels, to control plant diseases. The Chinese have been very successful in nearly eliminating certain seed-borne pathogens of wheat and rice by their strict adherence to principles of clean seed and quarantine. Fields with known disease or pest infestations are generally not used for seed. Flag smut was controlled in Shensi Province, at least in part, by importing seed from other provinces. Quarantine policies are probably excessive in some cases, e.g., a quarantine against *Tilletia controversa* Kuhn (TCK) in Kirin Province since 1966, from which time only spring wheat has been grown in this province (TCK is known only on winter wheat). Fields with take-all are also quarantined, and the seed from such fields may even be heat-treated before use. The take-all pathogen is not carried with seed, nor is the prevention of movement of this fungus from infested to clean fields an important consideration in take-all control (see the subsequent section on take-all). Other plant protection techniques applied to wheat are hot-water treatment of seed to control loose smut; benzene hexachloride on seed to control common bunt, or applied to soil for control of armyworms; salt water floatation to separate scabby seeds or nematode cockles from good seed; and organic phosphate insecticide (Rogor) used to control aphids and plant hoppers; and liberal use of organic manures (which have been recognized scientifically since the early 1920's for their suppressive effect on soil-borne plant pathogens).

The methods of disease and insect control used in China are obviously effective. Except for *Fusarium* head blight and associated scab in the lower and middle Yangtze River area, the wheat we saw in China was virtually free of diseases and insect pests. Individual specimens of many different wheat diseases and insect pests were found in the occasional field, but otherwise most of the wheat we saw was healthy (except possibly for nutrient deficiencies). This is a credit to the Chinese plant protectionists and a testament to the value of disease and insect control technology as it existed already 25-30 years ago.

One of the few research projects of a basic nature on wheat diseases observed by delegation members was at the Northwest Agricultural College in Wukung. This project was attempting to explain the origin of new races of stripe rust and concluded that a single-spored culture (single-spored only once, at the beginning of the test), if transferred through eight or more successive generations on a variety initially resistant, could develop virulence for the variety by the eighth generation. Either the single spore used to start the culture was heterozygous and carried a gene for virulence to the so-called resistant variety, which is mutated, or the culture was contaminated by a virulent race.

In Shanghai at the Institute of Plant Physiology, some 2,000 strains of actinomycetes and bacteria have been screened for antibiotic production that might be used against rice blast or wheat scab. The antibiotic, Qingfengmycin, has been identified from a strain of *Streptomyces aureous* and is effective against rice blast. (The Japanese have used antibiotics for years against rice blast in Japan, and Qingfengmycin may eventually find a similar place in rice production in China.) Another antibiotic, not yet identified, was said to be effective against scab, although we saw no tests. The project leader planned to provide cultures of the *Streptomyces* strain to the communes, together with a recipe for making a medium of wheat and rice brans and corn meal in jars so the communes could make their own antibiotic to control rice blast now and perhaps wheat scab in the future. We saw autoclaves on one commune (used to sterilize a medium for culture of a medicinal *Ganoderma*), but it was not clear how the communes were to maintain genetic purity of an antibiotic *Streptomyces* or standardize antibiotic production. The fungicide used presently for scab control is purchased through the county (see the section below on scab).

Testing for rust resistance was under way in virtually every breeding program visited at the provincial level. At some locations, rust inoculum was introduced by spreader rows. In addition, the program at Nanking inoculated one large block of lines with the scab fungus (*Gibberella zeae = Fusarium graminearum*) to test for resistance to scab. The group at Kungchuling in Kirin Province inoculated spring wheats with *Helminthosporium sativum* spores to test for resistance to leaf blight and crown rot caused by this fungus. There have been some tests for resistance to yellow dwarf at the academy at Wukung, and the group at Kungchuling indicated a strong interest in starting a testing program for resistance to yellow dwarf.

The monitoring of pathogen inoculum or insect buildup for the purpose of forecasting disease epidemics or insect problems on wheat is common in China but generally localized. Ascospore inoculum of *Gibberella zeae* is monitored by spore traps in the Shanghai area under the leadership of a plant protectionist at the Shanghai Academy of Agricultural Sciences. Aphids, leaf hoppers, and armyworms are monitored largely on the communes, for local information, but with some coordination by radiobroadcasting carried out at the county level. Rust epidemics are anticipated mostly by watching traditional sources of inoculum, e.g., mountainous spring wheat areas west of Shensi Province for stripe rust or more local pockets within the province where leaf, stem, or stripe rust usually appear earliest. The traditional hot spots for early rust development are apparently sprayed as necessary with *p*-amino benzoic acid to thwart possible epidemics (see below for more details).

DISEASES OF WHEAT

Head Blight and Scab

Head blight and the associated scab of the kernels, caused by *Gibberella zeae* (imperfect stage, *Fusarium graminearum*) is the most important disease of wheat in China and the only disease of economic importance not yet under control. This disease is particularly important in the middle and lower reaches of the Yangtze River but extends considerably north and south of this region, as well as into central China. The disease is favored by the warm moist weather that begins in April and early May. It is also favored by the double- and triple-cropping systems of wheat with corn or wheat with rice; the pathogen overwinters and eventually forms its *Gibberella* stage (the major source of inoculum) on the stems of both rice and corn.

The formation of perithecia on corn stalks is well-known. Their formation on rice stems is unusual for this fungus in nature but is very important in China. We were shown good evidence of the *Gibberella* stage of *F. graminearum* on rice stems, although the fungus apparently does little or no damage to the rice crop.

Although *F. graminearum* is the main cause of the head blight and scab in China, about 5 percent of the diseased samples yield either *F. culmorum* or *F. avenaceum*. The perfect stage of *F. avenaceum* has not been found in China, but the group at the Northwest Agricultural College at Wukung were familiar with Colin Booth's book on *Fusarium* taxonomy (1972) and knew from that book of the description of *Gibberella avenacea* R. J. Cook (imperfect stage = *F. avenaceum*) in Washington state.

The *F. graminearum* in China is probably of Group II proposed in Australia for this fungus, namely, the cause of head blight of wheat and stalk rot of corn, and capable of abundant perithecial production, even in pure culture. On the other hand, the occasional specimen of wheat crown rot and associated prematurity plant blight caused by *F. graminearum* was observed in Hopei Province, where wheat is double cropped with corn. These crown rot symptoms were similar to those associated with Group I of *F. graminearum* in Queensland, Australia, and the Pacific Northwest, United States. Efforts are under way through U.S. Plant Quarantine to import specimens of head blight and crown rot collected in China to determine whether they are of Group I or II, or possibly of a Group III, specialized for rice as well as wheat.

The most intensive screening program for resistance to scab was at the Kiangsu Provincial Agricultural Science Research Institute in Nanking. The workers at this institute culture *F. graminearum* on autoclaved barley seeds and then distribute these over the soil surface in March; perithecia develop on the kernels by early April and provide the ascospore inoculum needed to produce head blight in the nursery. Some effort aimed mainly at selection of wheats for resistance using naturally occurring inoculum was also underway at some of the communes in the Nanking and Shanghai areas, at the Academy of Agricultural Sciences in Shanghai, and at the Shensi Academy of Agricultural and Forestry Sciences at Wukung.

Several wheats showed markedly less scab compared to susceptible checks. Indeed, the Chinese may have better wheats for use under head blight and scab conditions than those of possibly any other country in the world. However, it seemed likely from our observations that the lack of scab in these wheats was a form of escape, rather than resistance, due either to (1) flowering while the head is still in the boot, (2) flowering within a tightly closed lemma and palea, or (3) flowering early, i.e., before appearance of the ascospore inoculum (which generally starts when temperatures reach 15°C). If so, then wheats "resistant" one year, or in one area, may be less resistant in another year, or another area, depending on whether inoculum production and the vulnerable flowering period happen to coincide. We saw good evidence to support this inference for barley as well as wheat. From this standpoint, the approach used by the Chinese to find improved wheats for the scab problem area is sound, namely, test the wheats only in the fields, and under the management conditions and in the areas where the variety ultimately will be grown.

The source of resistance to scab mentioned in the report of the Plant Studies Team, namely, Nung-lin no. 50 from Japan, was not mentioned by any of the breeders we met who work on scab. Instead, Frontana from Brazil apparently has been a popular source of resistance for the breeding program at Nanking. This may be no coincidence, since scab is also a major problem in Brazil. Varieties that appeared scab resistant at the Shensi Academy of Agricultural and Forestry Sciences, included an Albanian wheat (A-Po), one from Italy (Ch'eng Yin no. 1), a variety from Kiangsu Province (Su Chow), and a variety numbered 6701. The combined efforts of the various institutes, communes, and brigades will almost certainly produce an acceptable, high-yielding wheat for the scab areas within the near future.

In the absence of resistance in current wheats, fungicides are commonly used in the lower Yangtze River area. The wheat is sprayed once at about flowering and often a second time, 7 days later. All institutes and communes mentioned the same compound, a systemic fungicide sold by the state (through the county) and called "Bavistin" or "MBC". The structural formula was given as follows:

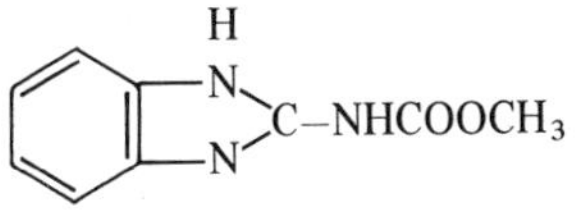

This compound, methyl 2-benzimidazolecarbamate, is the active component of methyl 1-(butylcarbamoyl)-2-benzimidazolecarbamate (benomyl). Benomyl is a systemic fungicide developed by Du Pont and used in the United States for the first time from the late 1960's to early 1970's. MBC is marketed under the name Bavistin by the BASF Company in West Germany, Australia, and India. It was not revealed whether the Bavistin used in China is purchased from BASF through one of the three outlets mentioned above or manufactured in China. We were only told that the compound is made available by the state. It is the only modern systemic fungicide we encountered in use in China, although apparently others are also available there and used as necessary.

The Plant Protection Lab of the Shanghai Academy of Agricultural Sciences has studied the time of ascospore release, using spore traps, in an effort to forecast scab epidemics and improve the timing for chemical application. Their spore traps are on many communes in the Shanghai area and are attended and read by plant protectionists at the brigade or team level on the respective communes. Four such spore traps were maintained on the Ma Lu People's Commune, which we visited. A count of seven ascospores in a microscopic field at 400× is considered the threshold inoculum level and time to apply the chemical.

Good drainage is used in the Shanghai area to reduce scab. Perithecial production is reduced by keeping the immediate soil surface as dry as is practical. Through the use of deep ditches and drainage tile, the water table in the Shanghai area is now maintained at 1 m or more below the soil surface of the wheat fields during spring and summer. In addition, the Ma Lu People's Commune applies Shanghai city garbage to its wheat fields during December to "keep the soil warmer"; if, indeed, this treatment helps raise soil temperatures during winter, it would also help keep the wheat growing and increase the chances of flower production ahead of peak inoculum production. The academy at Shanghai is experimenting with different planting dates which also affect time of flowering and, hence, disease potential.

A form of black chaff was noted on the heads of certain experimental lines at the Kiangsu Provincial Agricultural Science Research Institute in Nanking. This was probably physiological black chaff and was associated almost exclusively with CIMMYT wheats, or derivatives thereof. Physiological black chaff is a genetic trait commonly associated with certain CIMMYT wheats wherever they are grown.

Smuts

At least three smut diseases occur on wheat in China, namely, common bunt (*Tilletia caries* and *T. foetida*), loose smut (*Ustilago tritici*), and flag smut (*Urocystis tritici*). All were apparently quite important in the 1950's but are under control now, so far as we could determine. We found flag smut in rain-fed wheat near Wukung in Shensi Province. This province used to have fields with 1-10 percent flag smut, but we were told that the disease is now under control. We saw only mounted specimens of common bunt, including a specimen at the Kiangsu Provincial Agricultural Science Research Institute in Nanking labeled *T. foetida* and one at the Northwest Agricultural College at Wukung labeled *T. caries*. We were informed on several occasions that common bunt "no longer occurs in China." Loose smut was found in trace amounts in nearly every province visited but is apparently also under control.

Most smut control is apparently accomplished through seed treatments. At the Double Bridge People's Commune near Peking, a special team responsible for seed production treats the seed with hot water, for loose smut control, and also Cerasan (mercurial), presumably for common bunt. Mercurials are apparently still in use on the communes in China. At Ma Lu People's Commune, the seed is soaked in a solution of benzene hexachloride (666) or sometimes in limewater for several hours. The limewater treatment is an old method for control of the seed-borne phase of bacterial blight of rice and has apparently been adopted for use with wheat but with unknown or unproved benefits. This commune displayed on the wall large watercolor posters (approximately 18 × 24 inches) of various wheat diseases and disease cycles in their simply equipped plant protection lab. Loose smut was illustrated on one of the posters and also in various plant protection handbooks as a reminder of the constant threat this disease apparently represents.

Benzene hexachloride was also used in Shensi Province for flag smut control; seeds are soaked for 12 hours in a 0.3 percent solution. We were told also that ammonium chloride was applied in the past through the drill into the seed furrow for flag smut control. The extent to which this treatment is still in use is unknown. Its effectiveness and practicality, however, would seem questionable. Apparently, irrigation has also helped reduce the incidence of flag smut compared to that observed in the past on dryland wheat.

The seed treatments we encountered for smut control in China are old and have been long since replaced with more convenient or safer methods in many other countries. None of the plant protectionists questioned at either the commune or research institute level expressed an awareness of the breakthroughs of the past 10 years on systemic seed treatment fungicides for smut control. We also found no place in China where hexachlorobenzene (HCB) is used for bunt, although considering that a modern compound such as Bavistin is in use, HCB surely is also in use or has been in use on wheat in China. This highly effective bunticide has been in use in the United States, Australia, and other countries for about 20 years.

Tilletia controversa Kuhn (TCK), cause of dwarf bunt and the basis of a quarantine maintained by the Chinese against U.S. wheat since 1974 (when they found spores of TCK on wheat purchased for food), probably could not infect wheat in the vast majority of the wheat-producing area of China, even if introduced. The highly specialized conditions required by this fungus to infect wheat, namely, winter wheat covered 6 or more weeks with snow on unfrozen ground, simply do not occur in the winter wheat-growing areas of China. Most of China, especially the winter wheat areas, is very dry in winter. Even in northwestern Shensi Province and adjacent Kansu Province, snow lasts for only a few days, or sometimes up to 1-2 weeks. The upper reaches of the Yangtze River, e.g., Szechwan Basin and the plateau of Tibet, are winter wheat areas where possibly the winters are sufficiently mild for TCK infections and where the snow lasts. These areas are quite remote from the main areas of China.

When they were asked, "Where in China do the conditions occur that are required by TCK for infection of wheat?", our hosts answered, "We do not have TCK in China." We attempted to clarify the question by agreeing that TCK does not occur in China and then asking more directly, "Where *could* it establish if introduced accidentally?" The answer was, "The spores live a long time in the soil; we do not have TCK in China; we consider it a great danger to our wheat if introduced." We believe the Chinese concern for TCK is genuine, but we further believe that this and other experiences encountered in China suggest a general lack of appreciation among Chinese plant protectionists for ecology of plant pathogens and pests. They recognize the importance of summer moisture for rusts and other leaf diseases, but beyond this, they revealed little awareness of the importance of environment in disease development. Having discovered the quarantines imposed on wheat from province to province, even against seed from fields with take-all, having learned of the number of diseases controlled in China on rice, wheat, and other crops through establishment of rigid seed certification procedures, and having observed the great concern among the Chinese at all levels for clean seed, we see little hope that the Chinese will soften their stand against TCK spores on wheat from the United States.

Stripe Rust--*Puccinia striiformis* West

In the 1950's, stripe rust was an important disease in north and northwest China. The most severe losses occurred within 100 miles of the Wei and Hwang rivers from Sian to Chengchow and in the valleys north of the Hwai River. Other areas in which losses occurred in some years were western Szechwan Province in 1951, Inner Mongolia in 1952, central Hopei Province in 1953-55, and Kansu Province in 1957. There was apparently a local outbreak in southern Shensi Province as late as 1963. Stripe rust oversummers at altitudes of 2,000 m in eastern Kansu and Tsinghai, where maximum summer temperatures do not exceed 20°-22°C. The hosts in this area are probably late-planted spring wheat and volunteer wheat. The inoculum is carried southeastward in the fall by winds, with the rust overwintering on wheat growing along the river banks. During moist springs, stripe rust then spreads into Shensi, Shansi, and Honan provinces. The inoculum is returned to Kansu and Tsinghai provinces by winds from the east and southeast during the summer. It is not clear where the rust epidemics in Hopei Province originated, but it most likely was in the mountains south of the Hwai and/or the Yangtze rivers.

Stripe rust has been controlled by a combination of methods. Starting in the 1950's, Pi-Ma no. 1 and Northwest Agricultural College no. 6028 were introduced as the first major varieties resistant to stripe rust. The Chinese currently feel that they have 600 varieties resistant to stripe rust. Cultural control methods also are used. They include timely seeding, not too late in the mountains and not too early (10 days later than the previous normal) in the lower regions in the fall. We think that since the Cultural Revolution, the changes in cropping practices have had an even greater effect than that of resistant varieties on the control of rust in China. Double cropping in the lowlands has required the earliest possible harvest, and planting tends to be later owing to the time required for the second crop. Land leveling has been extensive in the lowlands, making overwintering on river banks and waste areas more unlikely. We saw little nonagricultural land in the crop-growing areas, so volunteer plants are probably rare. When stripe rust centers are found in the fall or winter, removal and burial of all infected leaves is recommended, followed by spraying of the surrounding area with a sodium para-aminobenzosulfate. This chemical is also used to control stripe rust outbreaks in the spring. In Hopei Province an additional chemical, zinc ethylene bisdithiocarbamate, is also used for stripe rust control. In Hopei we were told that chemicals are applied every 10 days once the disease prevalence reached 5 percent. During our trip we saw no stripe rust in production fields. Susceptible border and spreader rows artificially infected with stripe rust were observed in breeding nurseries at the Double Bridge People's Commune (Shuang Chiao) near Peking, the Kiangsu Provincial Agricultural Science Research Institute, the Shensi Province Academy of Agricultural and Forestry Sciences at Wukung, and the Hopei Cereal Crop Research Institute at Shihchiachuang. Only at the latter location was rust adequate for any meaningful selection for rust resistance.

Race identification of stripe rust for the People's Republic of China is done at the Shensi Academy of Agricultural and Forestry Sciences at Wukung. They use a Chinese set of differentials that include the following varieties: Northwest no. 54, Bumper Harvest (from Northwest Agricultural College), Pi-Ma no. 1, Strubes Dickkopf, Fulhard, and Trigo Eruoka. Race identification is done in a basement chamber at 9°-15°C, with 4,000 lux of light. Currently, there are 20 races in China, but only 6 are of any importance. Occasional local outbreaks of stripe rust have occurred in recent years when races changed.

The important race in Shensi Province is China race 20, while in Hopei Province it is China race 17. We were able to obtain both disease notes and variety names for only a few varieties observed in the Hopei Cereal Crop Research Institute. Blueboy had a trace of stripe rust; Scout, Scoutland, and Hyslop were resistant; and the susceptible varieties had 100 percent severity. Several laws about race shifts of stripe rust have been developed in Shensi Province:

1. Variation in pathogenicity occurs first in the west (oversummering area).
2. Variations normally occur on plants with a heavier than normal nitrogen application.

3. New races occur first on seedling plants, i.e., at the early stages of epidemic development. Pi-Ma no. 1 was released in 1951, a few infections were found on seedlings in 1954, and by 1957 some adult plants were rusted lightly.

Stripe rust is a disease of cool moist climates, with mountainous areas being notably affected. Thus, expansion of wheat into the mountainous regions of China will increase the hazard of stripe rust epidemics. Stripe rust in north and northwest China could be an annual problem only in the warmer and moist southern part. For stripe rust epidemics to develop throughout the rest of north and northwest China, it would require one or more of the following: (1) winters much warmer and wetter than normal, (2) a cool, wet spring, (3) additional inoculum in the fall from late-maturing wheat in the mountains, and (4) large amounts of volunteer wheat during July through September. Stripe rust would be a problem in northeast China if the inoculum source in north China were heavy and the summer were cooler than is normal, as was the case in 1950 when stripe rust was widespread in northeast China.

In our opinion, considerable work needs to be done on stripe rust epidemiology and control in north and northwest China, even though no rust was observed in those fields we visited in 1976. The questions that should be examined are the following: (1) How diverse is the resistance to stripe rust in the 600 varieties, i.e., how many different genes and how many different combinations of stripe rust resistance genes occur in these varieties? What is the distribution of these host varieties? (2) What is the current potential for a stripe rust epidemic from the vast changes in cultivation practices since the great Cultural Revolution? Where and how large are the inoculum sources for the fall-planted wheat? Do they vary in size and location from year to year? Is the lack of the disease in recent years due to (a) lack of inoculum, (b) unfavorable environmental conditions, or (c) resistant varieties? (3) How can an adequate level of stripe rust infection be obtained for selection in a breeding program? The level of infection must be a great deal higher (Hopei Cereal Crop Research Institute was an exception) than that we saw in 1976.

Leaf Rust--*Puccinia recondita* Rob. ex. Desm.

Leaf rust occurs throughout China on winter and spring wheats. In most of the production fields visited in north and northwest China we saw few infections; however, those we saw were of the susceptible type. The climate in the southern part of north China should be conducive to leaf rust epidemics, so the lack of disease in 1976 must be related to an absence of primary inoculum and/or possibly some varietal resistance. Elsewhere in north and northwest China, leaf rust should not be a major concern, since the dry falls and springs would generally retard leaf rust development. In those years when the summer rains come early, leaf rust could become heavy late in the season but result in only light losses. This condition would be reversed and leaf rust could be devastating if wheat were grown through the wet summer months or in those years with above-average winter-spring rainfall. In the spring wheat area of northeast China, leaf rust inoculum normally arrives too late (flowering stage of wheat development), and temperatures are generally too high for disease epidemics to occur.

Apparently, little emphasis is placed on breeding for leaf rust resistance; however, selection for leaf rust resistance is being done at the commune level. Many of the advanced lines in the breeding nurseries are more severely rusted than are the current commercial varieties. We saw some severe rust in a seed increase block of F_5 and F_6 material, indicating little effective early generation selection. No knowledge of the pathogen race distribution or host genotype for resistance was obtained for this disease.

We saw only parts of a nursery at the Wukung Institute of Agricultural Sciences and at the Crops Institute at Shihchiachuang where adequate leaf rust existed for good selection for disease resistance. At the Kirin Academy of Agricultural Sciences of Kungchuling we were told that they had a cold spring and therefore had little leaf rust; however, if the disease normally occurs in early June adequate rust should develop for selection for leaf rust resistance. Curiously, at both the Wukung and Shihchiachuang institutes, we saw the American varieties Scout, Scoutland, and Blueboy, which were resistant to leaf rust at those locations. At Shihchiachuang, Hyslop was also resistant.

In order to avoid possible losses due to leaf rust, it will be necessary to understand the environmental factors and the current host resistance(s) in use in the Yangtze and Hwai river valleys. This area appears to be conducive for leaf rust to overwinter, increase, and then spread northward into the spring wheat area in some years. An important control measure would be to assure that the spring wheats have at least two effective genes for leaf rust resistance different from those in the southern winter wheats and for which virulence is absent in the southern population.

Stem Rust--*Puccinia graminis* Pers.

Stem rust was not observed in any commercial fields we visited. Stem rust was severe in Inner Mongolia in 1956 and 1958, years when severe epidemics of stem rust occurred in the Yangtze and Hwai river valleys in the southern portion of the North China Plain. With the current planting and harvesting dates, stem rust probably will not be an important disease in north and northwest China. Insufficient hot weather and dews occur during April and May for disease development. Any extensive use of spring-planted grain would greatly alter this condition. Some interest was expressed in growing spring wheats in the Peking area. In northeast China, stem rust is considered a major hazard to the spring wheat crop. Stem rust was severe in northeast China in 1948, 1951, 1952, and 1956. Warm temperatures and frequent rains in this area from about the boot stage of growth until harvest make favorable conditions for stem rust development. The average maximum temperatures for July of 84°F (29°C) and 87°F (31°C) and minimums of 65°F (18°C) and 69°F (21°C) at Harbin and Shenyang, respectively, are ideal for stem rust, and rainfall is plentiful. Thus, all that is required for epidemic development are a susceptible host and adequate inoculum.

Although stem rust usually is not a problem in north China, the southern part of this area may be the source of inoculum for northeast China, as probably occurred in past epidemics. Increased acreages of wheat in south China could provide additional inoculum in the future to both north and northeast China.

Some physiologic race identification work is done at the Kirin Academy of Agricultural Sciences at Kungchuling. We were told that races 21 and 34, plus one other race, were the most common. The identification work is done in the greenhouse during the winter, so we did not see it. The formulas for effective/ineffective genes for resistance to these races would be as follows: race 21--*Sr* 5, 9e, 21/7b, 9d, 18, 19, 20, and 28; race 34--*Sr* 9e, 21/5, 7b, 9d, 16, 18, 19, 20, and 28. According to the literature, in 1958, races 21, 17, 34, and 40 were used in testing work. The formulas for these additional races would be as follows: race 17--*Sr* 5, 9e/7b, 9d, 18, 19, 20, 21, and 28; race 40--*Sr* 21/5, 7b, 9d, 9e, 16, 18, 19, 20, and 28.

The stability of the races over the years indicates the absence of a sexual cycle and probably a source area in which no selection pressure has been applied to the pathogen. Rust collections for the race survey are made by the technicians in the various production brigades in the provinces. Adult plants are tested for rust resistance in a field nursery. The technique used is to plant a spreader row of plants of which 50 percent are susceptible to stem rust and 50 percent susceptible to leaf rust. Stem rust-infected plants in pots are then transferred from the greenhouse to the field at a frequency of 1 pot per 10 m of row. Leaf rust is dusted on the spreader rows. At Kungchuling, stem rust normally occurs around June 10. We actually saw only two nurseries with stem rust infection, both of which were artificially inoculated, one at the Kiangsu Provincial Agricultural Science Research Institute in Nanking, and the other at Shensi Province Academy of Agricultural and Forestry Sciences at Wukung. In both, only traces of stem rust existed, insufficient for any type of selection. The spring wheat at Kungchuling was heading, and no stem rust was yet sporulating, which would indicate that the selection pressure there for stem rust resistance will also be low in 1976.

Apparently, after the stem rust epidemics in the 1950's, Thatcher (*Sr* 5) spring wheat from Minnesota was introduced. We were told that this variety was resistant from then until 1964, after which it was no longer grown. In the past few years the Chinese have started growing some Mexican and American spring wheats in their nurseries again. In one production brigade we saw a field of the CIMMYT semidwarf Tanori. In Kirin Province two major wheat varieties are grown, Feng Ch'an no. 2, which was released in 1959 by the Kirin Academy of Agricultural Sciences, on 60 percent of the acreage and Hsin Hsü Kuang no. 1, which was released from Heilungkiang Province in 1973, on 20 percent of the acreage. Thus, genetic diversity in this area is nil.

Powdery Mildew--*Erysiphe graminis* DC.

This was the second most common disease, but losses in the observed commercial fields generally were light or nil. We were asked several questions about sources of resistance; however, at the Hopei Cereal Crop Research Institute we saw a wheat line being increased for release that had powdery mildew so severe only the flag leaves remained green. The disease has become more severe with the increased use of nitrogen fertilizer and thicker stands.

Leaf Spots

Leaf spots are of little importance on wheat in most of China. Even in the lower reaches of the Yangtze River where leaf spots were most prevalent, it seemed doubtful that sufficient foliage was destroyed to reduce yields. However, several leaf spots and blights occur in China and, conceivably, in the right year, could cause damage.

Septoria tritici and *S. nodorum* both exist in China but are probably not as important as was suggested in the report of the Plant Studies Delegation. We saw no *Septoria* on wheat in China during our visit, nor was there concern or work on either species of this pathogen. An exception may be Kirin Province, where both *S. tritici* and *S. nodorum* have been found on spring wheat in the past. However, there is no work on either of these species.

Helminthosporium sativum also occurs on the spring wheat in Kirin Province to the extent that varietal plots are inoculated with conidia at about heading to screen for resistance. The specialists at the institute in Kungchuling refer to the disease caused by *H. sativum* in their province as "root rot," but in fact the pathogen causes a leaf sheath and stem blotch. Apparently, infections may be sufficiently severe to cause prematurity blight, although it was too early for us to see such blight.

Fusarium nivale causes a very conspicuous and somewhat destructive leaf blotch on wheat throughout the lower and middle Yangtze River area, and as far west as Wukung in Shensi Province. The blotches are elliptically shaped, 2-3 cm long by 1-1.5 cm wide, and are rustic red to brown. Infection is by ascospores from *Calonectria nivalis*. This sexual (inoculum) stage appears in profusion in the lower leaf sheaths of the wheat beginning in early April, or even late March, in the warmer areas. This is the same disease caused by *F. nivale* near Mexico City where CIMMYT wheats are developed. This leaf blotch is unique in the sense that *F. nivale* is best known as the cause of pink snow mold of winter wheats that overwinter for 3-4 months beneath snow on unfrozen ground. Apparently, China has one of the few habitats for this fungus where moisture and temperature are ideal for leaf infections, even in the absence of snow. The life cycle of this pathogen in China is also one of the few examples where the sexual stage plays a documented role in infection. It is the only place, to our knowledge, where functional perithecia of both *C. nivalis* and *Gibberella zeae* appear in mixed population and at about the same time.

Dilophospora leaf spot caused by *D. alopecuri* (Fr.) Fr. also occurs in China, according to recent Chinese literature, but we did not see this disease. Similarly, *Xanthomonas translucens* occurs in China but was not observed. There were numerous small spots, flecks, stripes, and blotches with no recognizable fruiting bodies of fungi and which thus could not be identified under the circumstances.

Root Diseases and Soil-Borne Plant Pathogens

Most, if not all, of the known soil-borne plant pathogens of wheat occur in China. On the other hand, root diseases caused by these pathogens are virtually nonexistent. This is almost certainly due to the liberal and regular use of organic manures applied as a source of fertilizer and for improving soil structure. It may also be due to the 3-4 month period of flooding for paddy each year; organic manures and flooding, together or singly, are excellent practices for biological control of root-infecting fungi.

Rhizoctonia solani Kuhn, cause of sheath blight and sharp eyespot, was abundant in Kiangsu Province on the lower leaf sheaths. However, the infections were characteristically superficial. This disease is of mycological but not economic interest.

Specimens of eyespot root rot caused by *Cercosporella herpotrichoides* Fron. were found near Sian in Shensi Province, and also near Shihchiachuang in Hopei Province. This is apparently the first record of *C. herpotrichoides* in China. This fungus can devastate wheat, if the conditions of prolonged, cool (5°-10°C), rainy weather are provided. It is probably significant that the specimens found were in the southern part of north China where rains were relatively frequent this past year, where winters are cool but mild, and where the fields are not used for paddy. Future epidemics are improbable, however, unless the use of overhead sprinkler irrigation for winter wheat increases.

Take-all, caused by *Gaeumannomyces graminis* (Sacc.) von Arx and Olivier var. *tritici* Walker (= *Ophiobolus graminis*), was observed only once (one plant in Hopei Province), although apparently it has been important in China in the past. We were told of recent outbreaks of take-all in western Shensi Province. The Shantung Academy of Agricultural Sciences has written a small handbook (dated 1974) for plant protectionists at the commune on diagnosis and control of this disease, suggesting that take-all may still be important in that province. The handbook has a full-page color illustration of various stages of take-all on wheat. Many recommendations for control are listed, but probably the most effective are (1) rotate with paddy rice at least one year; (2) incorporate about 5,000 kg of organic compost (straw, night soil, animal manure, and soil, all mixed together and fermented) per mu of land (= 75,000 kg/ha); (3) add 40-50 kg superphosphate/mu (= 600-750 kg/ha); and (4) turn the soil, that is, invert the surface 40-50 cm so as to bury the infested layer. Other recommendations that are probably less important or noneffective are (1) quarantine the infected fields; (2) heat-treat the seed at 51°-54°C for 10 minutes; or (3) treat the seed with a fungicide. *Gaeumannomyces graminis* var. *tritici* is not transmitted through the seed, nor are quarantines necessary, since the fungus is probably already on the wild grasses of most areas environmentally suitable to it and only needs the right soil conditions and two or more consecutive wheat crops to multiply and cause damage.

The near-universal practice of compost preparation with wheat straw (among other ingredients) could serve to introduce the take-all fungus in significant amounts if the straw came from a field severely affected by the disease. The Chinese are aware of this and eliminate the risk by fermenting the compost; such compost is stacked in piles 2-3 m high and 4 m wide, allowed to heat in the center to 50°C for 3 days, turned so the outer edges are in the center, and allowed to heat for 3 or more days. They also ferment the bran to destroy the fungus, but this is probably unnecessary.

The practice of one winter wheat crop every year with irrigation during the winter and spring sufficient to produce 5,000 to 7,000 kg wheat/ha would seem ideal for take-all; this only reinforces the conclusion that the disease is probably controlled biologically because of the organic amendments.

Fusarium root and crown rot occurs in the southern part of north China, but like other soil-borne diseases of wheat, is very rare. We were told in Shensi Province that *F. culmorum* occurred in some cases but that *F. graminearum* was more common. This is the only province where any kind of in-depth observation on the various diseases has been made by the local plant protectionists. Probably a wider search in the other areas where winter wheat is produced, dryland or irrigated and in rotation with corn, sorghum, or millet, would reveal more of this disease. The specimens of crown rot and associated prematurity blight caused by *F. graminearum* in Hopei Province were new to the plant protectionists accompanying us in that province.

In both the Pacific Northwest and Australia, *Fusarium* crown rot is most destructive in certain semidwarf wheats grown with ample or even surplus soil nitrogen. Similarly, in Hopei Province, the few specimens of this disease were found in the semidwarf wheats that had tillered profusely and were well supplied with nitrogen. This raises the possibility that parts of China could also experience more *Fusarium* crown rot as semidwarfs become more popular and more nitrogen is applied to reach and maintain high yields. The causal fungus is already well distributed in the area, as the cause of scab, and thus there is no shortage of inoculum.

Cephalosporium gramineum Misikado and Itaka, cause of the vascular stripe disease, apparently occurs in the western part of Shensi Province, according to plant protectionists at Wukung. We saw no specimens of this disease, but their description of it seemed accurate. Other root diseases that are present in China but that we did not see in the field were seedling blight caused by *Sclerotium rolfsii* (Sacc.) Curzi (probably in south China), root rot caused by *Helminthosporium sativum* Pam King and Bakke (probably in drier central China), and stem lesions caused by *Gibbellina cerealis* Pass.

Virus Diseases

There are at least five virus diseases of wheat in China, three of which apparently are not known to occur in the United States. The two common to both countries are barley yellow dwarf virus and wheat streak mosaic virus. The three present in China, but unknown in the United States, are rosette dwarf virus, Russian winter wheat mosaic virus, and blue dwarf virus.

Barley Yellow Dwarf Virus (BYDV)

This is probably the most important virus disease of wheat in China. Although we saw only sporadic occurrences of this disease during our visit, it has been important in the past. It was apparently very important in the Peking area in 1973. We saw it near Peking, Nanking/ Shanghai, and Sian in Shensi Province.

There are at least five species of aphids known as vectors of BYDV in China. The most important is *Toxoptera graminum* Rondani, which moves great distances by wind when in winged form. The virus oversummers (between wheat crops) on volunteer wheat seedlings and also on *Panicum miliaceum*, a common grass. Epidemics in Shensi Province are associated with the early-sown fall wheat (September) on the drier upland slopes to the west and north. This is the direction from which the prevailing winds blow in fall, winter, and spring, and thus the vector is carried southeast to the later-sown crops.

The plant protection division at the Shensi Academy of Agricultural and Forestry Sciences has five researchers assigned to work on virus diseases of wheat. They have examined about 1,000 varieties, and inoculated more than 200 artificially, to screen for resistance. As elsewhere, they have tolerance but no resistance. Tolerance has apparently been most evident in "native" wheats collected from the communes and presumably used by the peasants for many years; we were told that such local wheats are almost invariably more tolerant of BYDV than are wheats imported from abroad. If this is true, it would indicate that yellow dwarf is an old problem in this area of China and that much selection and wheat improvement has been done by the peasants through the years. The group at the Shensi Academy of Agricultural and Forestry Sciences claims to have about 20 wheats tolerant to BYDV.

Probably the main method of control of BYDV is by insecticides applied to control the vector. The communes are well equipped for spraying wheat with backpack sprayers, and these we saw in operation many times. In Shensi Province, Thimet (an organic phosphate) is commonly used. The even older but still popular benzene hexachloride ($C_6H_6Cl_6$) is also used.

The only cultural controls indicated were close planting and irrigation; both practices increase the density of the crop, which sometimes helps confine the aphids to smaller areas. In contrast to dense planting, the practice of space planting to allow for interplanting with corn or cotton has favored BYDV somewhat. This was especially evident in Hopei Province, where barley and wheat both showed more BYDV when planted as narrow strips (with space between for other crops) than when planted solid. This disease is commonly most prevalent on the edges of the field, and intercropping provides a nearly unlimited number of such edges. Continued use of intercropping will thus probably favor increased BYDV.

Rosette Dwarf Virus (RDV)

This is a problem mainly in northern China, including northern Shensi, Shansi, and Hopei provinces and Peking municipality. This virus is vectored by a plant hopper, *Delphacodes striatellus*. The most obvious symptoms include severe stunting and increased tillering, hence the name, rosette dwarf. Two names in local Chinese literature acquired and possibly synonymous with RDV are "rosette green mosaic" and "streaked dwarf virus." The plant protectionists on this commune knew it was vectored by a plant hopper. The leaves may develop white spots or streaks when young but become deep green when older. The leaf blades commonly are slightly to one side or the other, suggesting that one edge grows less than the other. The plants we saw were more chlorotic than white.

The virus apparently kills the growing tip; heads with sterile tips similar to that caused by frost are characteristic. Losses have been known to reach 30 percent in northern Shensi Province in the past.

The virus has been examined by electron microscopy at the Institute of Biochemistry in Shanghai. We saw a photograph of the particles; they were 5-6 times longer than wide and slightly wider at one end than the other. The tapered morphology may have been an artifact. Alternate hosts of the virus include dogstail (genus of millet), maize, and many other grasses. Blotches form on the maize, but it is not serious on maize. The virus does not carry through the eggs of the hopper.

Control is primarily insecticides to control the vector and persistence in removal of all infected plants. The Double Bridge People's Commune near Peking controls the disease with a rigorous spray program using benzene hexachloride.

Winter Wheat Mosaic Virus (WWMV)

This virus occurs in northern China in much the same areas as RDV. The virus is vectored by the leaf hopper (*Psamatettix striatus* L.). The disease is apparently the same as a winter wheat mosaic reported from Russia more than 30 years ago. Names used in local literature that may also be synonymous with WWMV are northern mosaic and northern cereal mosaic. We were unable to learn more details on the epidemiology or importance of this virus disease in China and presume the information is not available. We saw at least one probable specimen of this disease in a field near Peking.

Blue Dwarf Virus

This virus is also vectored by the plant hopper, *Psamatettix striatus*. It apparently occurs in the north and middle parts of Shensi Province and probably in the other provinces of northern China as well. We saw no confirmed specimen of this disease but were told that plants are stunted and that new leaves are white while old leaves develop a very deep green to blue/green color. We were informed that no work in China is under way as yet on this disease.

Wheat Streak Mosaic Virus

This virus occurs in winter wheat in the drier upland northern parts of Shensi Province and adjacent Kansu Province and possibly other areas as well. As elsewhere, the vector is the mite *Aceria tulipae*. Apparently, the mite oversummers on grasses, especially *Panicum miliaceum*, which grows during the wheat-free period of June through September. The disease has come under control since establishment of the People's Communes and a more organized and unified effort to separate wheat from the carrier grasses. The disease was described as sporadic at present.

Ergot and Cockles

Wheat cockles, caused by the nematode *Anguina tritici* (Steinbush) Filipjev, was not seen in the People's Republic of China. It has been a problem in the past but evidently is no longer. We were told that the diseased grains were removed from the seed by floating in saltwater, which is the same method currently used for scabby grain.

Ergot, caused by *Claviceps purpurea* (Fr.) Tul., probably has never been a serious disease of wheat in China. It has occurred in northeast China on winter rye. With advances in agriculture, rye has generally been replaced with spring wheat in the northeast, and thus, no ergot occurs. In triticale breeding, which is done mainly at Peking, they reportedly have not had much ergot. We looked at a commercial triticale seed production field near Peking and saw no ergot, even though some floral sterility existed. In the Peking area the normal spring seasons are probably too dry for floral infection by ergot.

FIELD INSECT PESTS OF WHEAT

The major hazard of insects on wheat in China is as vectors of viruses, namely, aphid species as vectors of barley yellow dwarf virus and plant hoppers as vectors of rosette dwarf virus, northern mosaic virus, and blue dwarf virus. These wheat viruses and the vector relationships have been discussed in the virus section.

Aphids also cause some direct damage to wheat in China. This was especially true in the lower and middle Yangtze River areas, where populations were very high and where honeydew in the area of stem-leaf attachment putrefied and killed the leaves or even the entire plant. Aphids were sometimes densely packed in some spikelets as well. In general, however, these situations were curiosities and not economic problems. Nevertheless, aphids were observed on the wheat at virtually all locations and are the number one insect concern of the Chinese. Moreover, the use of insecticides for aphid control represents the single greatest use of pesticides on wheat in China. It seems a fair estimate that 90-95 percent of the irrigated wheat, as well as a considerable portion of the dryland wheat, are sprayed at least once for aphid control. The most popular chemical for aphid control is Rogor (an organic phosphate), and the second most popular is probably benzene hexachloride. A good aphid control program in east-central China involved Rogor applied at least once during mid-March, and again in April or May. One application of Rogor includes about 2-3 kg formulation per hectare and costs about $1.50 per hectare.

The wheat blossom midge (*Sitodiplosis mosellana*) was a problem in the 1950's in some parts of China, but we were informed that current wheats are resistant to this pest. According to reports received at the Northwest Agricultural College at Wukung, resistance to blossom midge in wheats now grown in Shensi Province came from an early U.S. wheat, Quality. Although current wheats may indeed be resistant, the importance of changed cultural practices of the past 20 years and greater use of insecticides cannot be discounted.

Soil insects are the major insect pests of wheat in China. Armyworms are most important, but there is also sporadic damage from stem maggots and wireworms (*Pleoromus canaliculatus*). In Kiangsu Province, Diptherex was given as the most common insecticide used against armyworms. A commune in Hopei Province near Shihchiachuang reported to us that they control armyworms with 15-30 kg benzene hexachloride per ha, applied and mixed in the surface soil before seeding in the fall. This treatment costs about $0.75 per hectare.

As with the treatments for disease control, the chemical treatments used against insect pests of wheat represent technology of at least 20 years vintage. Only in Kirin Province, at the Academy of Agricultural Sciences at Kungchuling, did we encounter a wheat specialist familiar with a more modern insecticide, in this case the systemic Disyston, for aphid control. It was unclear whether the plant protectionists at this institute were only familiar with Disyston or if this insecticide is actually in use in Kirin Province. The Chinese are concerned with toxicity problems, however, and mentioned at most locations that Rogor should be used in preference to benzene hexachloride because it is less toxic to the people who apply the chemical. So far as we could determine, all insecticides and other pesticides and fungicides are applied by backpack applicator (or the equivalent), generally by several people moving abreast through the field, and often by people with cloth face masks for protection.

Biological control of the insect pests of wheat is of considerable interest in China, but has some way to go before it will replace chemicals even slightly. This area of endeavor may be one of the few where observations of a more basic nature are under way at the provincial institutes and academies. At the Kiangsu Provincial Agricultural Science

Research Institute in Nanking, several entomologists are gathering information on natural insect enemies of aphids and plant hoppers, with the objective of using these for biological control. At the Ma Lu People's Commune near Shanghai, the plant protectionists at the commune research station rear and release about 10,000 ladybugs and 6,000-7,000 adult lacewings per hectare to control aphids. Although the effort was admirable, our field observations indicated that it was also inadequate for the aphid population on wheat at that commune. At the Shensi Academy of Agricultural and Forestry Sciences at Wukung, there is apparently work on use of two soil fungi, *Metarrhizium anisophliae* and *Beauveria bassiana*, to control wireworms. These two fungi are well-known pathogens of soil insects. At the Soils Institute at Nanking, a study was under way on the morphology of the well-known insect pathogen *Bacillus thuringiensis*, using transmission electron microscopy (BM made entirely in Shanghai). We were given a photo they had produced of a flagellate cell. Whether or not *B. thuringiensis* is used in China for biological control of insect pests was not learned.

Other biological or cultural approaches to insect control more difficult to evaluate are of the "peasant lore" category. For example, a young plant protectionist on a commune near Kungchuling in Kirin Province stated that straw was placed on the soil surface to "trap" the eggs that give rise to armyworms; the straw was then removed and burned. This mixture of predominantly sound but some questionable approaches to disease and pest control on wheat in China is to be expected where plant protection is an art rather than a science and where the opinions of experienced peasants are probably more of a basis for decision making than is experimentation.

Nematodes

Seed cockle caused by *Anquina tritici* (Steinbush) Filipjev is apparently the only recognized nematode problem on wheat in China, and it has now been all but eliminated. It was apparently quite important in the past, but was brought under control by seed purification. Salt solutions of 20 percent sodium chloride were used to float out the cockles and separate them from the seed. There was also some importation of clean seed from areas free of the nematode into areas where the nematode was a problem. There is no record of the oat cyst nematode, *Heterodera avenae*, in China. The wheat/paddy/paddy rotation in the lower reaches of the Yangtze River seems like an ideal situation for free-living root lesion nematodes, but we had no means to investigate this possibility. We encountered no interest in nematodes of wheat or other crops at any of the institutes we visited.

SOIL MICROBIOLOGY

Like the other more basic sciences, soil microbiology aimed at an understanding of the microbiological processes in soil is virtually nonexistent in China, at least at the 11 institutes, academies, and colleges that we visited at the central and provincial levels. At the Soils Institute in Nanking, we were shown work on isolation of nitrogen-fixing bacteria on a nitrogen-free medium. The isolations were by soil dilution plating of paddy soil. The plates were incubated under aerobic conditions. We were also shown a simple respiration study with paddy soil in sealed jars at room temperature. The objectives of these studies were not clear. We suggested that the soil dilutions be heat-treated at 80°C for 30 minutes to separate spore formers from nonsporing bacteria and also that the plates be incubated anaerobically to pick up anaerobic nitrogen-fixing bacteria; this suggestion was much appreciated.

A project at the Institute of Botany of Academia Sinica in Peking has focused on nitrogen fixation on corn leaves by an *Azotobacter* sp. The only aspects of the study shared with us were the steps used to isolate, purify, and crystalize the nitrogenase enzyme from the *Azotobacter*. The chemistry of this project was sound and modern.

We encountered no work in soil microbiology related to wheat. The most obvious area where important questions remain unanswered concerns the decomposition and biological nitrogen transformation aspects in compost preparation and use. There is apparently no standard method for making compost except to use various proportions of straw, night soil, animal manure, and silt, generally fermented for at least 10 days. The fermentative action probably destroys most plant, animal, and human pathogens. However, it would be interesting to know whether some nitrogen fixation, e.g., by *Clostridium* species, also takes place during fermentation. *Clostridium butyricum* and *C. pasteurianum* are both known fixers of nitrogen under reduced conditions and with substrates such as wheat straw. Moreover, earlier work with these organisms by C.A. Parker in Australia showed that a small amount of clay added to the culture medium in flasks greatly increased the amount of nitrogen fixed. Conceivably, the clostridia could be introduced with the soil or the animal manures. It would also be valuable to know how much nitrification occurs as the result of aeration during turning and drying and how much denitrification occurs, if any. A detailed analysis and characterization of this system that has evolved through centuries of trial and error could provide considerable practical information.

4

WHEAT QUALITY, STORAGE, AND PROCESSING

WHEAT QUALITY

The philosophy of wheat production in the People's Republic of China can be summarized best by numerous statements to the effect that whereas in the United States the important considerations are yield, resistance, and quality, in the People's Republic of China the considerations are yield, resistance, and early maturity. Thus, it comes as no surprise that the team saw little work on wheat quality. Work on improving nutritional quality is limited and involves determination of total protein in agronomically advanced and promising lines. We learned about such limited work conducted at the Kiangsu Provincial Agricultural Science Research Institute, at the Hopei Cereal Crop Research Institute, and at the Kirin Academy of Agricultural Sciences. Only at the Hopei Cereal Crop Research Institute were foreign introductions evaluated; in the other two institutes, apparently, only Chinese wheats were studied. The team did not see or learn of any work on improving amino acid composition of, or balance in, wheat proteins.

Limited indirect work on improving end use properties is conducted in several of the communes and agricultural experiment stations. Tests include kernel weight and test weight (yield components that relate to flour yield in milling), kernel color, kernel texture (as it relates to protein content), and protein content for determination of nutritional value and bread-making quality. Red wheat types are chosen over white ones because of the susceptibility of the latter to sprouting and impaired end use properties. Primarily, the reason given for discontinuation of durum is its susceptibility to sprouting. While wheat selections are made on the basis of yield-resistance-maturity, it is of interest to note that the justification for developing octoploid triticale is its superior bread-making potential.

We learned of one instance only of breeding for improved milling and bread-making properties. The Albanian variety A-Po is grown widely in Shensi and Hopei because it has good disease resistance, a stable yield (on low-fertility soils), and late maturity (thus making possible good labor distribution). A-Po is susceptible to lodging, it produces grain with thick husks, the flour milled from it is of inferior color, and the gluten is of low quality and quantity. Crosses to produce a more desirable A-Po type were designed to eliminate lodging and improve milling and baking properties.

STORAGE

While decentralization and spreading of grain storage over a wide area is recommended for strategic reasons and to minimize transportation (by using the grain where it is produced), some storage is inevitable. We visited storage facilities in communes (mainly for their own reserves), in a granary in a municipality and/or county, and in a large flour mill.

Grain at the Chiang Ning Commune (near Nanking) is stored for as long as 2 years in 25-ton round bins built from straw and clay. The straw is mixed with clay and then twisted into thick cord-like pieces, which then are placed in the form of coils on the ground to form a circular structure. After drying in the sun, the silolike structures are strong enough to withstand the weight and pressure of the grain. The bins are constructed on rock-cement or lime-sandrock, 70-80 cm high foundations to protect the grain from ground moisture and rodents. Some of the grain at the Chiang Ning Commune is stored for as long as one year in 350-ton size warehouses.

About 4,500 tons of grain at the Ma Lu Commune (near Shanghai) were stored in small round storage bins and in a three-story storage building.

The Ta Hsing Lou granary near Sian was founded in 1951. It occupies an area of 6.4 ha and has 127 personnel. It is one of several municipal granaries-storehouses for grain from the communes in Shensi Province. Imported grain is stored elsewhere. The permanent capacity of the granary is 18,500 tons. Additionally, some small round bins are used. Permanent storage is contained in seven storehouses and 37 sheds. Storage capacity of each storehouse is about 1,500 tons divided into 14 compartments. Sheds range in capacity from 50 to 150 tons, round bins from 40 to 100 tons. All grain is brought into and out of the granary in sacks. Most of the grain in the storehouses was stored in bulk but supported by sacks on the sides and top. Some grain in storehouses was in stacks 14 sacks high, but grain in the sheds was stored only 9 sacks high. A sack of grain weighs 90 kg. The storehouses were made of concrete bricks. The round bins were similar to those in the communes. The granary had 33 belt conveyors and some mechanical equipment for loading and unloading sacks.

FIGURE 16 Straw mat-covered grain storage structures at the Ta Hsing Lou granary in Sian.

In addition to the Ta Hsing Lou granary, we saw grain stored in small, rectangular tarpaulin-covered sheds, in small round bins, and in caves along the railway (mainly near stations) on the way from Sian to Shih-chiachuang. Many round bins were seen on small vacant lots in the streets of Changchun. Some of the storage maintenance along the railway track seemed to be of somewhat inferior quality.

The Shanghai flour mill has 24 round storage bins each of 900 tons capacity and 14 smaller bins each of 200 tons capacity. Dimensions of the large round bins were 10 m × 20 m. The total capacity of the mill was 24,000 tons. The bins were constructed in 1936. Unloading from barges is by elevators. Work is being done on pneumatic unloading. Handling in the storage area was mechanized and centrally operated. Previously, 220 workdays (of 8 hours) were required to transport 1,000 tons to bins; now it is 25 days for 1,500 tons. Most grain is stored in bulk, some in sacks.

QUALITY OF STORED GRAIN

Grain delivered to the Ta Hsing Lou granary is inspected at the commune by a team composed of a brigade leader, brigade storekeeper, and representative of the peasants. Delivered grain must be dry, sound, clean, and insect free. The granary has cooling facilities for the incoming grain. Drying to about 12.5 percent moisture and treatment against insects is done at the commune. Grain condition is monitored by sensory evaluation, frequent checking of moisture, and thermocouples (85 per storehouse) for temperature changes. Inspected grain was dry, fairly clean (it contained some weeds), and insect free and had no odor. No signs of bird or rodent damage were evident. The granary maintenance and sanitation were excellent. Insistence on storing dry grain effectively eliminates microbial deterioration. Much of the grain is dried in air.

We saw grain dried on mats at the entrances of houses in cities, on balconies of apartments in communes, on parts of runways in airports, etc. Some mold-damaged grain was observed in the Nanking-Shanghai area where effective air drying was difficult. Properly protected and drained sheds and bins minimize bird and rodent damage. Minimizing insect damage is achieved, however, primarily by chemical treatment. To control insects in stored grain at the commune, one or two phosphine treatments per year may be used. The empty bins may be sprayed with benzene hexachloride (BHC) and aerated for 1-4 weeks before use. Still, we saw some insect-infested grain and, in one instance, production of tofu from very heavily insect-damaged beans. In general, control of insects is accomplished by careful inspection, monitoring, and good housekeeping combined with rather heavy chemical treatment.

Biological control is either very limited or in an experimental stage. It includes use of the fungi *Metarrhizium anisophliae* Metch. and *Beauveria bassiana* and of *Bacillus thuringiensis*.

Laboratories in granaries have large colored charts produced by the state for identification of the following insects: *Sitophilus granarius*, *Sitophilus oryzae*, *Rhizoperta dominica*, *Laemophilus minutus*, *Tribolium castaneum, Oryzaephilus surinamensis, Bruchus pisorum, Carpophilus dimiatus, Bruchtus chinensis, Araecerus fasciculatus, Trogoderma varium, Attagenus japonicus, Palorus ratzeburgi, Tenebrio obscurus, Ptinus fur, Lopocateres pusillus, Aphomia gularis, Plodia interpunctella, Sitotroga cerealella, Anagasta kuhniella, Aglossa dimidiata,* and *Pyralis farinalis*. We were told that the charts are for identification of insects from India.

Additional information on grain storage is of interest. No figures were available on economic losses in storage. Total shrinkage allowed by the state for losses in transportation, damage, etc., is 0.1 percent. We were informed that this shrinkage is adequate because the grain is transported (and stored in part) in sacks and is low in moisture when it is received in the granary. Grain turnover is about 50 percent per year. An objective is to store grain according to type and grade. This is generally impossible to implement, and much of the inspected grain was a mixture of soft white and red wheats. There is a small price difference among types and grades. Reused sacks are cleaned mechanically; there is no chemical or heat treatment.

PROCESSING OF WHEAT INTO FLOUR

It is our understanding that most brigades have a mill for custom processing peasants' grains or for processing wheat for manufacture of noodles for members of the commune or for the state.

Milling of wheat into flour was observed in five communes and in the Shanghai Flour Mill. The commune mills were in Shuang Chiao (near Peking), Chiang Ning (Nanking), Ma Chi Chai (Sian), Yung An, and Huai Ti (both near Shihchiachuang). The capacity of these mills ranged between 0.5 tons and 12 tons per 8-hour shift. The mills had very limited equipment for grain cleaning. Only one had a washer. Grain is milled dry. In one of the communes the 12 percent moisture grain was washed, and the washed grain was redried to 12.5-13 percent for low moisture in the stored flour. The milling equipment was Chinese, mainly from Tientsin. One of the mills used purifiers. In four of the five communes we saw roller mills, in one a disc mill. Only two products were produced: wheat flour and bran. The bran was generally reground in the mill. In custom milling, the bran may be returned to the households or sold to collective units. In milling for the state, the bran is sold to the collective units. The cost of milling in the Shuang Chiao (Peking) Commune was 0.6 fen/jin; the fee for milling was 0.9 fen/jin. In one of the communes, the household could get noodles from the factory in exchange for wheat. The flour extraction rate varied widely; it was 83 percent in the disc mill, and 83-92 percent in the roller mills. The number of workers per mill was generally 2-4.

The old, English-built, Shanghai Mill is one of the largest in China. The milling complex is comprised of three main parts (in addition to maintenance, repair, and construction): a grain storage and handling unit (built in 1936; total storage capacity 24,000 tons); a self-designed, self-constructed noodle factory (built in 1971; two noodle machines and one dryer; capacity 18 tons per 24 hours); and a flour mill. Workers numbered 1,600. The present milling output is 2,500 tons per 3 × 8 hour shifts. The mill has 93 rolls with a 17,780-cm milling length, i.e., 7.1 mm per 100 kg per 24 hours. This is an extremely short milling surface (a very short mill in the West would be at least 15 mm). The wheat (about 50 percent imported and 50 percent local from one state organization) is cleaned, washed, and conditioned (in mixed lots) for 16 hours. The conditioned wheat has about 13.5 percent moisture. The mill uses pneumatic transport for the mill stocks (only half installed at the time of visit) and mechanical transport for the wheat cleaning system. The present milling system includes 3 break + 3 middling with 2 additional rolls. The rolls operate at a differential of 2.5:1. This is a much shortened diagram compared to the original one (6B + 11M). Purifiers are no longer used.

It was claimed that shortening the diagram did not affect flour quality. This, however, is not substantiated by the fact that the permitted ash (dry matter basis) is 1.20 percent in the standard flour and 0.70 percent in the white flour. Power requirements were reduced by shortening the process from 60 to 34 kW/ton; moisture loss during milling was 0.3 percent. Two kinds of flour are produced: white (ching-pai), equivalent to 70 percent extraction (40 percent of total mill products), and standard dark (tiao-chun), equivalent to 80 percent extraction (30 percent of total milled products). In the production of white flour, 8-9 percent shorts also are obtained. The white flour should pass a 9XX sieve, the standard flour a 54 sieve, and the shorts a 42-wire sieve. The flour is bagged manually in 25-kg sacks. Flour is neither treated nor enriched; no additives or improvers are used. The white product is of an all-purpose type; the dark flour is used for manufacture of some bread and for processing into starch and gluten.

All wheat is milled according to a state plan, and all products are returned to the state for processing. The wheat is purchased at 290 yuan per ton; the cost of milled products (yuan per ton) is 400 for white flour, 330 for dark flour, 120 for shorts, and 80 for bran.

Samples of wheat and milled products were obtained. Yield, 1,000-kernel weight, ash, and protein of the People's Republic of China samples were compared with those milled experimentally on a rather simple mill from a U.S. hard red winter wheat at the U.S. Grain Marketing Research Center. The white flour in the Shanghai mill was definitely much darker in color and contained much more bran than the U.S. flour. (See Table 1.) The commune wheats varied widely in ash and protein; the ash contents of the commune flours ranged from 0.63 to 1.25, indicating a wide range in milling extraction and/or differences in effectiveness of flour/bran separation.

UTILIZATION OF MILLED PRODUCTS

Wheat is utilized in China in the form of yeast-leavened baked goods, noodles, and miscellaneous products such as ravioli, dumplings, pancakes, cakes, cookies, etc.

The yeast-leavened products are of two types: baked bread and mantou. Several members of the team visited the I-Li Foodstuffs Factory in

TABLE 1 Ash and Protein Content in Wheat and Milled Products from the People's Republic of China and a U.S. Experimental Mill

Product	Yield, %	Ash,[a] %	Protein,[a] % (N × 5.7)
United States[b]			
Wheat (27.2[c])	100	1.39	14.0
Flour	74.3	0.42	12.8
Shorts (26 ss)	2.1	2.64	18.0
Red dog (145 ss)	1.8	1.51	15.4
Fine bran (54 ss)	6.4	3.70	19.1
Coarse bran (26 w)	15.4	4.66	16.8
Shanghai Mill			
Wheat (37.1)	100	1.26	10.3
White flour	70	0.65	9.9
Standard flour	80	0.95	10.5
Shorts	9	3.75	11.9
Bran	20	4.08	11.9
Peking Commune			
Wheat (33.6)	100	1.55	10.7
Flour	85	0.81	10.9
Bran	15	5.47	14.8
Nanking Commune			
Wheat (33.6)	100	1.56	10.3
Flour	83	0.74	9.2
Bran	17	5.32	13.1
Sian Commune			
Wheat (35.4)	100	1.69	12.7
Flour	85	0.63	11.3
Bran	15	5.03	13.8
Shihchiachuang Commune			
Wheat (32.4)	100	1.51	11.9
Flour	88-92	1.12	11.5
Bran	8-12	5.09	13.8
Kirin Commune			
Wheat (33.9)	100	1.73	13.9
Flour	-	1.25	11.9
Bran	-	3.99	13.1

[a]Fourteen percent moisture basis.

[b]Experimental mill, U.S. Grain Marketing Research Center, Agricultural Research Service, U.S. Department of Agriculture, Manhattan, Kansas.

[c]Thousand-kernel weight, g.

Peking, which produces bread, biscuits, and candy. Types of other wheat products were observed mainly during meals. The I-Li Foodstuffs Factory in Peking produces in a single shift (10 p.m. to 6 a.m.) 10 tons of bread. The same factory also produces daily (in three shifts) 20 tons of biscuits and 25 tons of candy. Employees total 1,500. Most of the equipment in the factory is self-made. The raw materials for bread production are mainly from the neighboring provinces Hopei and Honan. The flour should contain 30 percent gluten for bread and below 25 percent for biscuit production. They must meet general specifications of the Foodstuffs Department of the Peking municipality.

The raw materials are stored in sacks. Bread is produced in batches of 250 kg of flour mixed with 112.5 kg of water for 10-12 minutes (depending on flour strength). Other ingredients (per 250 kg of flour) include 0.5-3.5 kg of salt, 12.5-37.5 kg of sugar, 1.25-3.75 kg of yeast cake, and 3.5-10 kg of peanut oil. No milk products or malt are used; lard is occasionally added in small amounts; some products are enriched with vitamin B_2 and $CaHPO_4$ (as a source of Ca).

Bread is produced by the sponge method: 1/3 + 2/3. Fermentation in each stage is 4 hours and is followed by a 40-minute proof. The bread is baked 7-30 minutes at 180°-220°C in a three-stage tunnel oven. In addition to the mixer and oven, the bakery has large fermentation vats, a molder, and a cutter.

The following main bread types are marketed: 100-, 150-, or 250-g panned plain; 150-g tien-yuan mien pac (sweet round), or fruit bread (with nuts and plums). Bread is freshly consumed; most of it is wrapped; practically all is from white flour; very little is toasted. Crumbs of freshly baked bread should contain 40-45 percent moisture depending on the type.

Team members were served on several occasions Pullman-type bread. The bread was tasty if freshly consumed. We saw different types of bread (about 250-g loaves, panned, on the way to the Great Wall; 150-g flat loaves in the Shihchiachuang railway station) being sold by vendors. Other types of baked or fried products included rolls, filled bread, fried-twisted bread, porous rolled out bread, and corn-soy bread.

Mantou is a yeast-leavened product produced from a stiff water-flour dough with little yeast, fermented at room temperature for about 3 hours on a damp cloth, scaled, and molded into pieces of varying size (25-100 g in general), and steamed for 20 minutes. The final product has no crust. There are many modifications of the homemade mantou with regard to dough formulation, size, shape, and preparation. Hua-chuan are steamed rolls which resemble mantou, multilayered with oil between the layers, generally not sweetened.

The team saw production of noodles in the Chiang Ning People's Commune (Nanking) and in the Shanghai Flour Mill. In both, the noodles were produced from a wheat flour (rather than semolina) milled from a rather soft wheat. In the Chiang Ning Commune, four workers engage in the production of 1.5 tons per 8 hours. The locally produced flour is relatively dark. A stiff dough (20 kg of water plus 80 kg of flour and 1 kg of salt) is sheeted, cut, and dried to about 10 percent moisture. The capacity of the Shanghai mill noodle factory was 18 tons per 24 hours. The equipment included two self-designed and manufactured machines and one dryer. The noodle machines have automated delivery and cutting of dry noodles; drying is in a traveling band oven for 2 hours at 50°C. One-half kg packages, 20 cm long, are handled manually. In the commune, the noodles were produced from a stiff dough of 100 kg of white flour plus 14 kg of water. The moisture of the dried noodles is 12 percent maximum. The price of a 1/2-kg package is 33 fen; there is a 3-fen price differential between wholesale and retail.

We were served noodles in different forms--in soup, as a main dish, and as chin-ssu-chuan (golden bread rolls). The latter are prepared as a product from noodles in an oiled dough, steamed, and cut into 10-15 cm long and 2-3 cm wide pieces. It is impossible to list all the types (or even main types) of miscellaneous wheat base foods. In addition to pancakes, cakes, biscuits, cookies, ravioli, and dumplings, the following should be mentioned: gluten balls--fried and often served with vegetables; chiao-tzu, a special type of dumplings, which can be steamed (most of the time), boiled, or fried; two types of fried, unleavened cakes, lao-piang (multilayered, 10 × 5 cm) and pao-piang (2-layered, can be taken apart and filled with meat, etc.); and finally some convenience foods, such as instant noodles, rolls with eggs and cream, etc.

NUTRITIONAL ASPECTS OF WHEAT AND WHEAT PRODUCTS

The average allocated amount of grain in the 10 communes that were visited ranged from 440 to 480 lb per capita per year. Under the assumption that all the allocated grain is used as food with an extraction of 85 percent, the yearly flour consumption would be about 385 lb, compared to 110-125 lb in North America. Rice continues to be the main grain staple, but consumption of wheat is on the increase and is encouraged by the state for agricultural and economic reasons. Wheat may comprise 10 to 50 percent (or even more) of the total grain. Rice continues to be the main and prestige grain in the south.

Protein content of the wheat is generally low (about 12 percent on a dry matter basis). Very limited work is done on increasing the protein content of wheat or improving its nutritional (rather than functional) value. No work, apparently, is done on improving amino acid balance of wheat proteins.

The average cost of food is about one third of total expenditures in the commune; it is presumably much higher in cities but no precise figures could be obtained.

LABORATORY FACILITIES, METHODS, AND SERVICES

We visited several laboratories that provide analytical assistance to plant breeders, processors, and technical personnel in granaries. The laboratories were simple and contained little equipment; practically all was made in China.

The laboratory in the Shanghai mill had equipment for subdividing wheat samples (a Boerner sampler and divider), standard sieves for ground mill products, ovens for moisture determination, an ash muffle furnace, and a gluten washer. The laboratory conducts no tests on the wheat. The wheat must meet state requirements as it is provided to the mill. No experimental milling or baking tests are run on incoming wheat. All tests are run on the milled stocks. We were told on two occasions about limited experimental milling and baking tests in the San Chiao Tsun production brigade (Yung An People's Commune) near Shihchiachuang and in the Sino-Albanian Friendship People's Commune outside Peking. The experimental testing facilities were not among the places visited. Tests conducted in the Shanghai mill included test weight (kilogram per hectoliter), moisture, ash (640°C), gluten, colorimetric tests for Fe and Pb, sieving tests (9XX for white flour, 54 for dark flour, and 42 for shorts), and Pekar tests of the flour (made every hour for visual comparison with a state sample). The I-Li food factory in Peking had a small laboratory for determination of moisture and gluten in all flours and free phosphates and heavy metals in biscuit products. In addition, the workshop conducted sensory tests and limited evaluation of physical properties of the food. It was our understanding that moisture was the only test occasionally run on Chinese noodles produced in the two factories.

We visited several laboratories that assist plant breeders in their selection. We were told that some of the testing (kernel size, color, and hardness) is done in the field by the plant breeders. The laboratory tests included determinations of sugars in plant tissue and protein in the grain.

We saw laboratories in a relatively small granary in the Chiang Ning People's Commune near Nanking and the large Ta Hsing Lou municipal granary in Sian. The main equipment in the granary included balances, drying ovens for moisture determination, an electric mill grinder, and an electric rice dehuller (with two adjustable rubber rolls) to determine the amount of husks. The laboratories in the municipal granary were housed in three rooms. One contained balances, air ovens, equipment for test weight determination, and grinders. In the second one were muffles, colorimeters, and equipment for extraction and determination of sugars, fat, and fatty acids. The third room had equipment for TLC, a distillation unit, a centrifuge, and some auxiliary equipment. This was the largest and best-equipped laboratory that we visited. In addition to frequent determinations of moisture, the laboratory followed changes in storage by determining sugars (total and reducing) and free fatty acids. The TLC equipment and the colorimeters were used to determine residual and adventitious pesticides.

To the extent that we could determine, there are large variations in methodology and there is no standard book of methods. The methods of the Association of Official Analytical Chemists (AOAC) are apparently consulted in the preparation of mimeographed instruction sheets. We saw such sheets in some laboratories, but in others, the instruction is by word of mouth and notebooks. The standard method for moisture determination is apparently 4 hours at 105°C, but we saw many variations: 8 hours at 100°C, 20 minutes at 160°C, etc. The general factor used for conversion from Kjeldahl-N to protein is 6.25; for wheat, 5.7; for triticale, 6.0.

GRAIN QUALITY AND STANDARDS

The best and cleaned grain is generally sold to the state immediately after harvest. At the storehouse in the Chiang-Ning People's Commune near Nanking the stored wheat was required to contain less than 12.5 percent moisture, 1.5 percent impurities (sand, soil, and weed seeds), 5 percent scabby wheat, and no live insects. However, a limited (and unspecified) amount of damaged grain (insects, heat, etc.) was allowed.

Scabby wheat was separated by flotation, both for use as food and seed. For seed use, a solution of 10 kg of salt per 50 kg of water was used; the above salt concentration is adequate to float an egg one third above the surface. The grain, from which scabby kernels were removed, is dried immediately. For use as food, flotation is in plain water. We were informed on several occasions that ergot (*Claviceps purpurea*) and the field insects *Aelia* and *Eurygaster* are not a problem in the People's Republic of China.

Weeds of major concern include: chickweed, foxtail, bindweed, goosefoot, and wild oats. The standards for wheat in the Shensi Province (according to the Ta Hsing Lou granary near Sian) are the following:

Grade	Kilograms per Hectoliter (Minimum)
I	75.0
II	73.0
III	71.0
IV	69.0

Total maximum impurities are 1 percent, and maximum moisture, 12.5 percent; the wheat was required to be free of scab and live insects. According to the information obtained at the Shanghai mill, the desirable moisture content of wheat was about 12 percent.

Wheat for the state was graded according to the following scheme:

Grade	Kilograms per Hectoliter (Minimum)	Total Impurities (Maximum), %
I	79.0	0.4
II	77.0	0.8
III	75.0	1.2
IV	73.0	1.6

The wheat had to be sound. No infested, scabby, or otherwise diseased wheat was accepted.

Local wheat milled for bread making should have at least 30 percent wet gluten, for other uses about 26 percent. This requirement cannot be implemented, however, in all cases.

Little, if any, testing was done on state wheat as it arrived at a granary or mill; the wheat was inspected at the commune before it was shipped.

The price difference between wheats from various grades is small

(13.1 to 13.6 fen/jin). While there is a good market for spring wheat, little is moved from province to province to meet quality requirements because minimizing transportation is desired.

5

POLITICS AND POPULISM: IMPLICATIONS FOR WHEAT PRODUCTION, RESEARCH, AND EXTENSION

The Chinese discern two major turning points in the recent history of China. The first turning point was "Liberation" in 1949. Liberation marked the end of the old, evil society and the beginning of a new era. The contrast between the pre- and post-1949 periods is stark. Before Liberation, material conditions were bad, and political rule was repressive and exploitative; since Liberation, material conditions have rapidly improved, and political rule has been by and for "the people." The second turning point, currently stressed with virtually the same emphasis as Liberation, was 1966 and the beginning of the Great Proletarian Cultural Revolution. Whether the Cultural Revolution warrants this heavy emphasis may be questioned. The Chinese themselves, however, aver that the salient characteristics of their government and society today have emerged as a result of the struggles, begun in the Cultural Revolution, between the correct revolutionary line of Mao Tse-tung and the revisionist line of Liu Shao-ch'i, Lin Piao, and Teng Hsiao-p'ing.

The current contention is that Liu Shao-ch'i and Teng Hsiao-p'ing, respectively vice-chairman and secretary-general of the Communist Party before 1966, were betraying the revolution during the years 1949-66 and were taking Chinese society on the path to the reestablishment of capitalist class domination. The Cultural Revolution aborted that rightist-deviationist attempt and restored the dictatorship of the proletariat. The struggle against those who would betray the revolution has, however, persisted. It is contended that Lin Piao, who had been a leader of the Cultural Revolution and had then been identified with the extreme left-wing radicals, was similarly a rightist-deviationist working to restore capitalist domination over the proletariat. In 1974-75, Teng Hsiao-p'ing repented his previous crimes, declared himself in support of the proletarian revolution, and was restored to a place of high authority. But Teng soon resumed his previous errors. His renewed betrayal of the revolution was discovered shortly after Chou En-lai's death in January 1976, however, and he again became the target of nationwide vilification. The delegation was in China in the midst of this national campaign and heard the rhetoric, saw the *ta-tzu-pao* (wall posters), and listened to the songs, plays, and movies that heaped calumny upon this renegade. Despite the reputed success of these successive struggles against the capitalist roaders, the major publications of the party (*Hung-ch'i*, *Jen-min jih-pao*, and *Chieh-fang jih-pao*) in May proclaimed that a "large portion" of the members of the party and state apparatus are still endeavoring to subvert the proletarian struggle.

What were the results of the Cultural Revolution? What is at stake in these continuing political struggles? From the vantage point afforded the delegation, it appeared that the Cultural Revolution had politicized society and mobilized the masses to a degree that was unprecedented before 1966. And it appeared that the continuing political struggles resulted, in large part at least, from differences over the question of whether these policies of politicization and populism should be perpetuated.

The uniformity and pervasiveness of politicization that we observed surpassed our expectations. When we visited an institute, college, factory, or commune, the format of a meeting seldom varied. We would be escorted into a sizeable reception room with a large table in the middle and chairs around the edges of the room. At one end of the room, there would be a portrait of Chairman Mao. At the other end of the room, there would usually be four portraits, those of Marx, Engels, Lenin, and Stalin. Quotations of the Chairman would be hung on the other walls. Served tea and cigarettes, the chairman or vice-chairman of the local revolutionary committee would identify his organization, give a brief account of its history, and remark about its achievements since Liberation. At this point, the political rhetoric was invariably introduced. Liu Shao-ch'i's crimes were named, and then it would be asserted that as a result of following the correct revolutionary line of Mao Tse-tung and the Great Proletarian Cultural Revolution, the revolution had been restored to its proper course and the achievements of the organization since 1966 had been greater then ever. And now, it was asserted, as a

FIGURE 17 Briefing of the U.S. Wheat Studies Delegation by the Chinese hosts in Peking.

result of the current campaign against Teng Hsiao-p'ing, the socialist consciousness of the people had been even further enhanced. The achievements now are thus unprecedented, although there still remain many shortcomings and there are still many things to be accomplished.

The introductions given by institution personnel did differ slightly from one organization to the next. The political atmosphere at Northwest Agricultural College in Wukung, Shensi (about 90 km west of Sian), for example, was especially thick and conformed with our understanding that China's educational institutions are peculiarly subject to political pressures from radical elements in the state leadership. The high level of politicization was, however, evident at all levels of the system with which we had contact. Old laborers in factories, young heads of production brigades, heads of scientific institutes, and rusticated youth in the communes--everywhere we visited--voiced the same phrases and gave identical interpretations of recent events. And not only in formal situations, but also in chance encounters with individuals, the same rhetoric could be heard. The agricultural technician at the commune near Shihchiachuang, the waitress in the Peking Hotel, the interpreter in Shanghai, and the guide at the Big Goose Pagoda in Sian--all, in friendly conversation and without our encouragement, uttered the accustomed phrases.

The Chinese acquire this rhetoric as a result of constant reiteration. Small-group study (hsüeh-hsi) sessions, involving some 10 to 20 members in each group, had been meeting during the campaign against Teng Hsiao-p'ing about twice a week for a total of approximately 6 hours a week. Songs, newspapers, radios, and wall posters all carried the same message. And from these several sources, the Chinese acquire not only the rhetoric but a world view that to us, at least, seemed to be thoroughly uniform and politicized.

Significantly, the system does not conceal anomalies or disguise that which is difficult to explain. Instead, the anomalies are confronted head-on, and the people are provided full rationalizations for that which might otherwise create doubts and misgivings. The revolutionary betrayals by their former leaders--Liu Shao-ch'i, Lin Piao, and Teng Hsiao-p'ing, for instance--have been thoroughly discussed in the press and in their small-group sessions. And the Chinese discuss these questions articulately and with seeming conviction. The Chinese with whom we discussed these matters (these did not include common farmers or laborers) are also imbued with an all-encompassing world view and prepared to discuss in some detail such diverse issues as the probability of war with Russia, the economic crisis in the West, and the race problem in the United States. Political ideology has thus penetrated deeply into society, and we felt--although this, of course, is an extremely difficult factor to judge--that this had been accomplished with remarkable success.

MANUAL LABOR

Perhaps the most far-reaching effects of the Cultural Revolution, however, have resulted from the extraordinary stress on populism--on the policy of eliminating elitist governance and attitudes and of fostering the active participation of "the people" in all areas of national life. The goals of the policy of populism are dramatized in Mao Tse-tung's goal of eliminating the Three Great Differences: the differences between city and countryside, between mental and manual labor, and between worker and peasant. To abolish the distinction between mental and manual labor, all persons whose occupations do not involve them directly and continuously in manual labor, except the elderly and physically unfit, must periodically engage in *lao-tung*, or manual labor. Administrators, teachers, and researchers must all lao-tung. The means by which this policy is implemented vary markedly throughout the system. We were told in Hopei, for example, that administrators (*kan-pu*) at the commune level must lao-tung for a total of 200 days each year; at the *hsien* (county) level they must lao-tung for 100 days out of the year. At the Sino-Albanian Friendship Commune, which is part of the municipality of Peking, by contrast, cadres at the commune level lao-tung for only 100 days each year; but cadres at the production brigade level must lao-tung for 200 days a year.

At the higher levels of the system--the prefectural (*ti-ch'ü*), provincial, and central administrations--the pattern of performing lao-tung is different. Generally, these cadres are expected to attend a May Seventh School for cadres once every 3 to 4 years for a period ranging from about 6 months to a year. The system, however, apparently allows for considerable flexibility. One central government cadre, who first attended a May Seventh School in 1968, has returned twice since then for a year each time. Another central government cadre, however, last attended the cadre school in 1970, when he was there for nearly 2 years. In their governmental offices, the cadres customarily take turns in going to the schools, but this practice of rotation may be altered if the work they are performing is particularly pressing. The pattern may also differ from province to province, for, we were told, some provinces are less forceful than others in enforcing the ideological remolding of their cadres.

Intellectuals, researchers and university professors, must similarly engage in manual labor and political re-education. At the Kirin Provincial Academy of Agricultural Sciences, for example, researchers and technicians, during any 4-year period, spend 6 to 12 months in a May Seventh School and 1 year on a commune engaging directly in production work there. In addition, the researchers assist with the manual work relating to cultivation of the institute's experimental fields. This pattern of manual work seemed generally to characterize all the institutions that the delegation visited, although again there were variations. At the Shanghai Academy of Agricultural Sciences, the researchers and cadre attended a May Seventh School for 2 to 3 months each year. In addition, usually on Thursdays, they work in the institute's fields, tilling the crops, digging new irrigation ditches, repairing roads, etc.

THE OPEN-DOOR APPROACH TO SCIENTIFIC RESEARCH

The populist policy was especially evident to the delegation in the intimate relationship that had been forged between the scientists and the tillers, in what the Chinese call "conducting scientific research in an open-door way" (*k'ai-men pan k'o-yen*). This approach dictates that scientific research must be conducted through the constant and close cooperation of scientists and peasants. Thus, the peasants may benefit from the theoretical insights and research achievements of the scientists, and the scientists are aided by the peasants' experience, practical understanding, and awareness of the people's problems.

The basic structure of agricultural research and extension work in the nation consists of seven levels. Organizations at four of these levels are operated by the state, viz., by the central government, provinces, prefectures, and hsien. The bottom three levels, communes, production brigades, and production teams, are operated by the people collectively. All levels, with the occasional exception of production teams, maintain experimental plots.

Most theoretical study and research is conducted by institutes at the state level, particularly, we would judge, at the central government and provincial levels (the delegation did not visit institutes or experimental plots at the prefectural or county levels). At the Hopei Provincial Institute of Agricultural Crops, for example, there are 230 employees, 40 percent of whom are directly involved in research. The institute has an experimental plot of over 60 ha. The Shanghai Academy of Agricultural Sciences has 1,404 employees, 480 of whom are research workers and scientists. And the Kirin Provincial Academy of Agricultural Sciences employs 805 staff members, of whom 259 are researchers, 436 are ordinary workers, and 110 are administrative workers. The Kirin institute has 100 ha devoted to experimentation, and over 300 ha are used for feed crops and seed.

The four lower levels of the structure are linked together in what the Chinese term the "four-level scientific and technical network" (*ssu-chi k'o-hsüeh chi-shu wang*). At the top level of this network, the hsien maintains a research institute (*nung-k'o-so*). This may be a sizeable organization. Near Shihchiachuang, for instance, Cheng Ting Hsien maintains an experimental farm, comprising 25 ha and 34 scientific researchers. It does work on soil and fertilizers, seed breeding, crop cultivation, plant production, and animal husbandry. At each commune, there is a scientific-technical station (*k'o-chi chan*) with anywhere from five to 10 persons with specialties corresponding to those at the hsien level. Then, in each production brigade, there is a scientific-technical corps (*tui*) with three to 10 members. And, in each production team, there is a scientific-technical group (*tsu*), usually with three to five technicians engaged in experimental work. The terminology applied to these several levels differs somewhat in different parts of the country, but the basic structure is uniform. Under this system the amount of land thus devoted to experimentation and the number of persons directly involved in scientific work are enormous. In the Ma Chi Chai Commune near Sian (with a total population of 22,200 persons and cultivated land of 1,800 ha), for example, 540 persons are involved, full- or part-time, in agricultural research, and 50 ha of land are devoted to agricultural experimentation and seed production. In Huai Te Hsien in Kirin Province, 12,304 peasant technicians are involved with agricultural experimentation at the production team level. In the nation as a whole, an official in the Agricultural Association told us, there is a total of 10 million persons involved in agricultural research at all levels.

The Chinese place extraordinary stress upon linking these several levels together. Before Liberation, they told us repeatedly, scientific researchers isolated themselves in their laboratories, unresponsive to peasant needs and incapable of communicating new scientific knowledge and techniques to the peasants.

One means by which knowledge is now disseminated through the structure is by periodic meetings. In Hopei, for example, a meeting of agricultural technicians and scientists is held once a year. And, at the hsien level, meetings are convened twice a year, for 3 to 5 days each, attended by about 800 persons representing provincial, prefectural, and county research institutes, as well as peasant technicians from the commune and production brigade levels.

In accordance with the open-door approach, scientific personnel spend much of their time working in the fields and with the peasants, while veteran peasants are also invited to work in the research institutes. At most research institutes, one third of the research staff, and sometimes two thirds of the staff, are absent from their home institutes because they are doing research with and working with peasant technicians at the commune and lower levels. Provincial institutes, as a result, frequently have a deserted appearance, with laboratories locked up and scientific instruments covered with plastic sheets.

Scientists remain at the "grass roots" for varying lengths of time. Seed breeders and planters of the Kirin provincial institute, for example, spend approximately 6 months of each year working with and exchanging information with peasant technicians in the communes within their area of jurisdiction. Other specialists, such as soil and fertilizer specialists, may spend only 2 months each year away from the institute. This research work at the local levels is distinct from the periodic stints of manual labor required of each scientist as part of his political education.

The various state-run research institutes also maintain "research bases" (*chi-tien*) at the commune and production brigade levels. The function of these research bases is essentially that of the scientific-technical stations maintained by the communes, except that the level of research conducted in the research bases tends to be slightly more advanced and theoretical. The numbers of research bases maintained by the various institutes differ greatly. The Hopei provincial institute, for example, operates only eight research bases. The Shanghai Academy of Agricultural Sciences maintains 38 research bases, in which 138 persons, 11 percent of the institute's total work force, are employed.

The scientists who are sent to work on the research bases remain at the local level for anywhere from 1 to 4 years. During that time, they live in the homes of peasants, sometimes eating with the peasant families but usually eating in the commune mess halls. The researchers' families remain in their homes near the mother institutes, the researchers returning to visit their families occasionally, especially during the slack seasons on the communes.

One such researcher, from the Hopei provincial institute, was interviewed at the Chih Ma Commune, some 45 minutes from Shihchiachuang. He stated that he expected to remain at the commune for 2 to 3 years. Then, he would return to the institute for approximately 2 years, after which period he would again be sent to a research base at the commune level. Most of his time at the research base is devoted to agricultural labor, he said, but he also engages in research on the research base's experimental plot and has extensive contacts with peasant technicians on the commune. One might expect such a scientist, separated from his mother institute, from the city, and from his family, to express misgivings about the system. To the contrary, he gave every evidence of being genuinely enthusiastic about the program. He declared that in the period before the Cultural Revolution, agricultural innovations spread to the production front very slowly if at all. Now, new information is disseminated rapidly and widely so that the fruit of research at the institutes is quickly put into practice by the peasants. In addition, he said, the system benefitted him because now he has a practical understanding of agricultural production and of the problems of the peasants, whereas previously his knowledge had remained largely theoretical.

Another aspect of the open-door approach is that experienced and skilled peasants are sent to work at the research institutes. Usually, these peasants remain there for 2 to 3 years, after which they return to their homes and other veteran peasants are brought in. At the Shanghai Academy of Agricultural Sciences, however, the peasants were not rotated, and they could remain at the institute "for a long time." These peasants, like the scientists at the research bases, leave their families at home, visiting them periodically during the year. There seems to be no uniform policy among the several institutes regarding the number of veteran peasants that should be brought into an institute. At Shanghai, the institute had only three peasants in residence. The Institute of Genetics near Peking, however, had 18. Researchers and administrators were loud in their praise of these veteran peasants. They assert that the peasants assist in research by bringing to the institutes their knowledge of farming techniques, irrigation, etc. They are, moreover, especially active revolutionaries and therefore help the institutes maintain their revolutionary vigilance.

THREE-IN-ONE COMBINATIONS

The government's determination to break down the barriers between intellectuals and manual laborers, which is a fundamental goal of the open-door policy, is evident also in the so-called three-in-one (*san-chieh-ho*) combinations. All administrative organizations in China--from the Central Committee of the Communist Party, the Chinese Academy of Sciences, and the research institutes and universities to the communes, production brigades, and scientific-technical teams--are organized according to the principle of three-in-one combinations. The basic purposes of this system are to prevent elitist, authoritarian rule; to guarantee that all constituent groups are represented on their leading bodies; and to guarantee the continuous training of future leaders of the revolution. The methods of attaining these purposes are, however, flexible and diverse. The definition of the three-in-one combinations that the delegation encountered most frequently was based upon age. According to the criterion of age, approximately one third of the personnel in a leadership group should be elders (*lao-nien*); one third should be middle-aged (*chung-nien*); and one third should be youth (*ch'ing-nien*). The youngest members of such combinations whom we encountered were in their middle and late twenties. Elders were those 55 years old and above. Generally, we found that persons in their thirties and forties tended to bulk somewhat larger than one third among the leading administrative personnel. Interestingly, both the party constitution of 1973 and the national constitution of 1975 stipulate that leading organs at all levels must include representatives of the three age groups.

The other chief criterion for selecting leading personnel in three-in-one combinations is "function." In most research institutes, for example, members of the revolutionary committees are selected from cadres, scientists, and workers. In most of the communes and production brigades that the delegation visited, the revolutionary committees comprise cadres, veteran peasants, and scientific-technical workers. Frequently, the criterion of age is applied along with the criterion of function.

Not always is the formula of the three-in-one combination limited to only three constituent elements. At the Sino-Albanian Friendship Commune, for example, there are nine distinct categories used in applying the so-called three-in-one combination to form the commune's revolutionary committee. Age is the principal criterion, but leading cadres, middle-ranking cadres (from the production brigade and production team levels), women, educated youth, mass representatives, militiamen, cultural-educational representatives, and factory workers were also included. The three-in-one concept is implemented with great flexibility, our source stated, because the basic purpose of the system, aside from training revolutionary successors, is simply to ensure that all major groups in an organization are represented in the leadership structure.

The three-in-one combinations assume alternative forms in other organizations. At Northwest Agricultural College, for example, the revolutionary committee of the college is formed of teachers, students, and cadres. A seed selection committee at the Hopei Provincial Institute of Crop Studies, is formed of one cadre, two peasants, and seven scientific-technical workers. A group in a production brigade that is responsible for selecting grain to be sent to the state is composed of the brigade leader, the chief storekeeper, and representatives of the poor and middle peasants. And the revolutionary committee of a block of workers' apartments is formed of teachers, parents, and retired workers. Whatever the method of forming the three-in-one combinations, the Chinese boast, the three-in-one combination is the backbone of production and a major source of their success.

REVOLUTIONARY COMMITTEES

The leading personnel of virtually all administrative units that the delegation visited--institutes, communes, production brigades, a college, factories, and a neighborhood committee (but not production teams)--are organized into revolutionary committees. There is, however, considerable diversity in the composition, structure, and operation of the different revolutionary committees.

In the communes, for example, the revolutionary committees are customarily selected by representative congresses (*tai-piao-hui* or *she-yüan wei-yüan-hui*), which are theoretically the highest authority in each commune. These congresses are made up of elected representatives, usually elected by representative congresses of the production brigades. In one commune, there is one representative for about every 40 persons; in another, 20 to 30 persons elect one representative. The representatives' terms of office vary from commune to commune. In one, a new congress is elected every 2 years; in another, every 4 years. The congresses also meet with varying frequency. In one, the congress meets once or twice a year; in another, the congress meets approximately every month, though less frequently during the busy seasons. All major questions affecting the commune, such as crops to be raised and capital construction to be undertaken, are discussed in the congresses. It may be doubted that these representative bodies are effective decision-making bodies, however, for they tend to be large, often numbering 800 or more members.

Once during the term of office of a commune representative congress, the representatives will select members of the commune's revolutionary committee. Usually, it seems, the congresses elect the members of their revolutionary committees by secret ballot. At the Sino-Albanian Friendship Commune near Peking, however, the selection process is considerably more complex. First, the incumbent revolutionary committee recommends a slate of persons to the representative congress. The representatives then bring the names of the nominees to the masses, who discuss the qualifications of the nominees and who in some cases propose alternative names. The representative congress then discusses the nominees again, after which it presents its recommendations to the outgoing revolutionary committee for final decision.

Variant methods of selecting revolutionary committees were found in other administrative units. At the Huai Ti production brigade near Shihchiachuang, there is a representative congress made up of one delegate for every 10 persons in the brigade. This congress meets only once a year. In some other production brigades visited by the delegation, there is no representative congress, and the revolutionary committees might then be elected by a secret ballot of all members of the brigade. In some units, the revolutionary committees are selected not by election but through discussion. In a Shanghai street committee, for example, the residents discuss the qualifications of potential revolutionary committee members in their small-group sessions. This is apparently a thorough and elaborately structured discussion; we were told that the discussion "goes up and down, up and down," until the discussion culminates in a "consensus." Other production brigades and the research institutes frequently employ this same method of discussion-and-consensus in selecting their revolutionary committees.

The revolutionary committees vary considerably in size. One commune, with 32,000 persons, for example, has a revolutionary committee of 37 members. Another, with a population of over 28,000, has 29 members, while a commune with 19,000 members has a revolutionary committee numbering only 20. At the production brigade level, revolutionary committees vary from six to 15.

Each revolutionary committee is headed by a chairman and often two vice-chairmen. In the larger revolutionary committees, such as those at the commune level, there is also a standing committee (*ch'ang-wei-hui*), which assumes responsibility for the day-to-day operations of the communes. Only a part of the revolutionary committee's membership is engaged full-time in administrative duties. In a revolutionary committee with 29 members, for example, 20 are regularly engaged in production work and only nine are full-time administrators. In one production brigade, the revolutionary committee counts 11 members, of whom seven are engaged directly in production, while only four devote much, but not all, of their time to administration. In a provincial research institute, the revolutionary committee has 33 members, a standing committee of 13, and five full-time administrators.

None of the revolutionary committees that the delegation visited, with but one exception, is headed by a woman. Several of the vice-chairmen of commune revolutionary committees, however, are women, and we were informed of production brigades that are headed by women. Women are, indeed, usually represented on the various revolutionary committees, although by no means in proportion to their total numbers. The revolutionary committee of the Huai Ti production brigade is typical: of 15 members on the revolutionary committee, two are women. The one exception to this pattern that the delegation encountered is the revolutionary committee of Tsao Yang New Village, a block of workers' apartments on the edge of Shanghai. Here, 12 members of the revolutionary committee are women, and only six are men. The chairman and leading members of the revolutionary committee are also women.

Party members seem to dominate the revolutionary committees. The delegation did not examine this topic systematically, but in the two revolutionary committees where answers to this question were obtained, the results were striking: in a revolutionary committee of 29, 22 are party members; in another consisting of 18 members, 13 are party members.

RUSTICATED YOUTH

Of the several methods being employed to eliminate the Three Great Differences, none contains a greater potential for political explosiveness than that of sending educated city youth to live and work in the rural communes. This program was initiated before the Cultural Revolution, the purpose being to reduce the size of the bulging cities and to raise the educational level of the villages. Youth from Shanghai were being sent to live in the countryside at least as early as 1962.

Since the Cultural Revolution, however, the program has gained momentum. Eleven million city youth, we were told, have been sent to the rural areas since the Cultural Revolution, 1 million of them from Shanghai. In one province, Kirin, there are 400,000 rusticated youth. Most of these city youth are sent to live on communes near their native cities rather than to the remote provinces. At the Ma Lu Commune, some 50 km from Shanghai, for example, the youth are overwhelmingly from Shanghai. We found the same situation near Sian, Shihchiachuang, and Changchun. At the Erhshihchiatzü Commune in Kirin, the youth came largely from nearby Kungchuling (a city of about 100,000 persons), and from Ssuping and Changchun (both about 60 km away). Of the 442 youth in the commune, 32 came from Shanghai and 13 came from provinces other than Kirin. These had been sent there, we were assured, only if the youth especially requested to be sent to the more remote, border provinces.

Youth who are sent to the countryside ostensibly do so voluntarily, although leaders of the revolutionary committee of a Shanghai street committee did state that parents often had to persuade their children to volunteer. And the move to the countryside is for both parents and the educated youth, initially at least, an obviously wrenching experience. A father of a rusticated girl from Shanghai related that he and his wife had initially been concerned for the health of their daughter and feared that she could not withstand the arduous labor on the farm. Later, however, by his account, they saw that their daughter was healthier and more robust than ever before and that she was happy in her work. As a consequence, he professed to be pleased that the daughter is living in a commune. One rusticated youth candidly admitted that he had initially objected to leaving the city. Owing to the invidious influences of Liu Shao-ch'i and Lin Piao, he said, he had wanted to continue his education so that he could become an official. Now, however, he realizes how wrong his attitude at that time had been.

Another youth, who had been sent to the countryside in 1968, gave this account of his experience and of the significance of the movement to send educated youth to the countryside:

> I was 19 when I came to the village. I volunteered to come, but initially, of course, I encountered many difficulties. Previously, living in Shanghai, my parents had always cared for me, and I was not accustomed to the hard manual labor in the village. We educated youth saw, however, that the peasants are greatly concerned for our welfare. We also saw that it was the peasants themselves who were transforming the countryside and the nation. Moreover, we realized that a major question confronting the nation and threatening the maintenance of the dictatorship of the proletariat is the irrational distribution of the nation's population. We youth understand very clearly that it is precisely because there exists such a large difference between the cities and the villages that we must go to the countryside, thereby removing excess population from the city and raising the cultural level and production in the village. As the economy develops, some of us will continue our education, and some will remain in the villages. But we educated youth have a saying that we will be happy to do

either, because in a socialist country all those who work are contributing to the task of socialist reconstruction. And, though we formerly wanted to attend a university, we now realize that the countryside is also a university.

The delegation spent the greater part of one morning visiting a group of educated youth in the Erhshihchiatzu Commune in Kirin Province. This commune has acquired a regional fame as a result of its rusticated youth. The youth there are housed in "collective households" (*chi-t'i-hu*), each household containing 20-25 persons. The household that we visited, with 10 boys and 12 girls, is an integral part of a production team. The youth work on the production team's fields, till their own "private plots," do their own cooking, wash their own clothes, and perform the other chores that help make their household a virtually self-sufficient living unit.

The significance of the educated youth in the communes is disproportionate to their numbers. Although the majority of them are engaged full-time and directly in farm production, others are serving as skilled workers (electricians, lathe operators, etc.), barefoot doctors, school teachers, nursery attendants, and accountants. They often take a leading role in political discussions. Several of the youth have become leaders of production teams or responsible members of production brigade and commune leadership groups. And these youth are especially active in organizing spare time "cultural teams," such as film projection units, musical and theatrical groups, and athletic teams.

Most of these rusticated youth, we were told, expect to spend the rest of their lives on their communes. A sizeable percentage of them, however, are transferred out of these communes and assigned to different duties. At the Erhshihchiatzu Commune, for example, where 443 educated youth remain, 418 youth have gone elsewhere since 1968. Of these, 270 have been sent to work in factories, 63 have entered a university, and 85 have been enlisted into the army. At other communes, the rate of exodus is lower. Outside of Sian, the Ma Chi Chai Commune retains 130 rusticated youth of the total of 200 who have been sent there. Elsewhere, the authorities were less precise and insisted that the great majority of the city youth remain in the commune. At the Ma Lu Commune near Shanghai, we were even told that, of 600 rusticated youth, all have remained in the commune.

IMPLICATIONS

The Chinese constantly reiterated to the delegation that China's great advances in wheat production are attributable to the correct revolutionary line of Chairman Mao. And the delegation agreed, after a month's visit, that wheat production is indeed deeply affected by the political policies that are manifestations of the Maoist revolutionary line. Are those effects upon wheat production, however, positive or negative?

The problems attacked by the Chinese are real enough. China's cultural traditions and behavioral patterns were deeply ingrained. These were the traditions that had made China's society one of the oldest and most stable in human history. They were, however, traditions suited to a relatively unchanging agrarian society in which a narrow stratum of relatively highly educated scholar-officials dominated over a broad mass of untutored peasants and manual laborers. In this traditional culture, a distinct set of behavioral patterns was implanted. It was a highly elitist society, in which the ruling class avoided manual labor and denigrated those who worked with their hands. The lettered elite tended to be bookish philosophers and moralists rather than practical technicians or natural scientists. Even the political administrators shunned contact with the people, whom they ruled and taxed. And the commoners, for their part, feared their social and political masters, were obsequious toward them, and tilled the land diligently but with little scientific knowledge.

Maoist policies are designed to shatter these cultural and behavioral patterns. Not only are the Maoists determined to eliminate political and economic exploitation, but they are convinced that production can greatly improve only if the functional divisions between scientists and producers are obliterated. Thus, scientific researchers must not remain shuttered in their laboratories, but must, as a result of living and working as farmers, familiarize themselves with the farmers' problems and empathize with the farmers' struggles. And the farmers must enhance their technical understanding by obtaining a better education, by participating directly in scientific research, and by sharing in the decision-making process. This, in brief, is the reasoning that underlies the policies of lao-tung, the open-door approach, the three-in-one combinations, revolutionary committees, and sending educated youth to the countryside.

That these policies have fundamentally altered the lives and methods of China's wheat scientists and farmers is indisputable. The Chinese farmer still works hard and long, but his life is no longer the impoverished and uncertain existence of the years before the revolution. And the scientist is no longer the tender-handed, ivory tower pedant of yesteryear.

The Wheat Studies Delegation was deeply and favorably impressed by the results of most of these policies. The farmers appeared to be well fed, healthy, and reasonably content; the political leadership seemed to be intelligent, informed, and committed; the scientists appeared to be vigorous and enthusiastic. And the production of wheat, as discussed elsewhere in this volume, evoked the hearty, if sometimes qualified, praise of the delegation's scientific personnel.

In only two respects did the delegation sense that the policies discussed in this section might have significant negative effects. There was the overweening stress upon *current* production problems and *practical* research, and there was the truly *radical* implementation of the open-door approach to management of science and research. The delegation recognized that the Chinese have created what is probably the most effective system of agricultural extension to be found anywhere in the Third World. Every aspect of wheat production is designed to provide an interchange between the farmers and scientists. New scientific knowledge and production techniques are disseminated quickly and effectively to the farmers in the fields. Conversely, the farmers can promptly inform the higher-level scientific personnel about difficulties encountered or notable achievements registered in their work. We felt that research and extension workers in other developing countries could greatly benefit by learning how the Chinese communicate scientific technology to the farmer. What, however, will be the long-term effects upon wheat research, the delegation wondered, of the frequent interruptions of the scientists' inquiries resulting from long periods of lao-tung and prolonged visitations at the commune level. The delegation also wondered if the politics of popular involvement in scientific research was not blunting the cutting edge of theoretical inquiry and was not dissipating the energies of the scientific personnel to a degree not compensated by the positive effects of a more effective extension system. These questions are examined in greater detail elsewhere in this volume. Here, however, it may be remarked that, in general, the delegation members tended to withhold judgment on these issues. They recognized that the Chinese currently assign exceedingly high priority to the production of food grains, and when China can be confident that it can adequately feed all of its population at all times, then Chinese scientists may be able to devote greater and more sustained attention to long-range, theoretical research.

6

EDUCATIONAL PROGRAMS AND MANPOWER TRAINING

The Wheat Studies Delegation had the opportunity to visit a variety of research institutes and colleges administered by the Chinese Academy of Sciences, the Chinese Academy of Agricultural and Forestry Sciences, and several provinces. Since the visits were invariably short, the delegation was able to make only the most cursory observations; however, by talking with scientists, teachers, and scientific personnel who accompanied the delegation, some generalizations can be made.

Chinese education is still in the throes of reorganization, a process that was set in motion by the Cultural Revolution some 10 years ago. Undergraduate education has been cut back from the pre-Cultural Revolution standard of 5 years to 3 or 3 1/2 years. The curriculum has been closely tied to short-term, practical objectives, which in agriculture means producing higher and more stable yields rather than taking long-term, theoretical subjects. Access to postsecondary education is limited to those candidates who have completed at least 2 years of work on a commune or in a factory or service in the army. In addition, candidates must possess credentials achieved by a high degree of political consciousness and motivation as well as the correct class background. It must be felt by the candidate's commune, factory, or army unit that the unit will directly benefit from the higher education that the candidate will bring back to the group. The preparation of teaching faculty and researchers for institutes and colleges seems to be achieved by extending the period of college training into what might be best called an apprenticeship program.

It is clear that the Cultural Revolution's effect on the Chinese educational system has left many institute and college faculty unsure as to the long-term trends in such areas as curriculum development. Most of the faculty met were teaching from lecture notes that they had written themselves. Another effect of the Cultural Revolution, the decentralization of the research process, meant that several of the institutes visited were conducting virtually identical research programs, although often not knowing of progress at the other institutes.

Members of the Wheat Studies Delegation made a variety of more finely focused observations of the education in certain disciplinary areas. Although the observations do not pretend to be inclusive, they are useful as a comment on some aspects of Chinese agricultural education and manpower training.

WHEAT BREEDING AND GENETICS TRAINING

The level of academic training varies greatly among the men and women wheat breeders in the People's Republic of China. There are many scientists with advanced training from foreign as well as Chinese universities. These people are very capable scientists and are providing the scientific basis for wheat breeding as it is now being conducted in the People's Republic of China. It would appear that a high percentage of those currently engaged in wheat breeding are in the 20- to 40-year age group. For these people, their training in plant breeding is mainly the result of practical experience gained by working closely with the older scientists. Some have had college training; however, it was not always in the biological sciences. Others attended technological school or were high school graduates who had been assigned to various communes, brigades, and production teams. Training for these people has resulted from scientists visiting the various communes where seminars and classes are presented and by learning from the farmers who have gained a great deal of practical experience. One of the real joys in the People's Republic of China was to hold discussions with the so-called educated youth of the commune or brigade level regarding various aspects of plant breeding. In addition to being extremely dedicated, they were very bright and gifted young people who seemed never to run out of questions or comments about their work. It would be most enjoyable and challenging to have such young people in our colleges and graduate programs.

Discussions were held with faculty members at the Northwest College in Wukung. Since 1966 the curriculum and other aspects of the college have been uncertain. As a result, it was not possible to obtain much information regarding the type of educational training that was offered. However, the focus on practical considerations appeared to receive the greatest attention. Chemistry was learned as a part of studying soil fertility and chemical seed treatments. Genetics was a part of agronomy. The duration of the programs was 1, 2, or 3 years, with all graduates returning to their respective communes upon completion of their program. Much of the training appeared to be field oriented. Again, the need to solve immediate problems received emphasis. With the older scientists nearing the end of their professional careers, one must wonder where the well-trained scientists with sufficient training in the basic sciences will emerge to provide the balance that will be so vital to the People's Republic of China in future years. As yield levels are increased and with the intensity of crop production, there will be problems emerging that will require well-trained scientists to provide solutions.

PLANT PATHOLOGY AND ENTOMOLOGY TRAINING

Plant pathology and entomology are pursued in China for the sake of plant protection, or they do not exist. As the name implies, plant protectionists for wheat devote their time to the art of keeping wheat healthy and not to the development of basic information about the disease and insect problems of wheat in China. A plant protectionist may be a technician with some training, a peasant with experience and possibly some training, or a specialist with considerable experience and training at a provincial institute or research academy.

At the production brigade level, there are plant protection teams. They are trained to recognize common diseases, plan spray schedules, and observe differences between varieties. The commune also has a plant protection team. Their makeup varies; however, at this level some members may have some biological training, perhaps as little as junior or senior middle school. However, some communes have school teachers, institute or college personnel, and students for a period of field work. The communes we visited normally had the same information available concerning disease epidemiology, control, and recognition as that of the institutes and colleges we visited. Some have posters (excellent) or collections of insects and diseased specimens to aid in identification. Some communes have microscopes. The personnel at the brigade and commune level may occasionally visit the province-level academies, institutes, and colleges for a field day.

Training and knowledge in plant pathology and entomology was also lacking somewhat in the provincial institutes and academies, but like the communes, there was no shortage of interest and enthusiasm for the subjects. The best entomology group for insects of wheat was at the Kiangsu Provincial Agricultural Science Research Institute in Nanking, and the next best was at the Northwest Agricultural College at Wukung in Shensi Province. It struck us that these two groups possess the majority of all information on insect pests of wheat in China. The best plant pathology group for diseases of wheat was at the Shensi Academy of Agricultural and Forestry Sciences in Wukung, where 12 specialists (including five on rusts and five on wheat viruses) concentrate on wheat diseases. There was at least one plant pathologist at each of the other locations: Kiangsu Provincial Agricultural Science Research Institute in Nanking, the Shanghai Academy of Agricultural and Forestry Sciences in Shanghai, the Northwest Agricultural College at Wukung, and the Kirin Institute of Agricultural Sciences at Kungchuling. These plant pathologists were nearly all trained before 1949.

PLANT PHYSIOLOGY TRAINING

The only opportunity we had to observe training of new plant physiologists was at the Northwest Agricultural College at Wukung. There we met with a group of seven plant physiologists, three of whom were older and obviously well trained. One was a student and three were younger staff who said little during the 2-hour session. In response to a question, it was stressed that education in plant physiology must involve practical problems in how to grow plants and serve the masses. We asked who was doing the teaching, and apparently six of the seven were doing some teaching. It involved both classroom and practical experience on the communes. We asked who would take over for the three older staff members (a plant biochemist, a botanist, and a plant physiologist), and we were told that others would be trained. We pointed out that they were getting near what we understood to be retirement age in China and that someone would be replacing them soon. The responsible person (a younger man) indicated that the student who was present was being trained for that. (She was a girl whom we would judge to be in her late twenties.) In her response to our questions she indicated that she had had no classroom training in plant physiology *per se*, biochemistry, chemistry, or statistics. The responsible person broke into the discussion and pointed out that she was learning more valuable things from her work on the communes. She was studying or had studied general botany in the classroom. When we asked if it would not be important to replace retiring staff with persons trained in biochemistry, chemistry, statistics, etc., the answer was, "Of course, and these will be provided." We did not encounter evidence that they were being provided at the agricultural college, however, but we did meet young researchers at some of the institutes who obviously had received basic science training.

The students whom we met at the Northwest Agricultural College had received no training in experimental design in their programs, and we saw little evidence that the active researchers in crop culture had received any. This appears to be an expensive omission in the Chinese agricultural curricula.

QUALITY CONTROL TRAINING

To the best of our knowledge, the training of technicians and technologists is done almost entirely on the job by experienced workers with some input from university/agricultural college personnel.

We had the opportunity to obtain some details while we visited the Ta Hsing Lou municipal grain storage unit in Sian, the Shanghai Flour Mill, and the I-Li Foodstuffs Factory in Peking.

Training at the municipal grain storage unit was by experienced personnel with considerable input from the Northwest Agricultural College and limited input from others. Experienced workers at the unit engage in continuous training of commune and brigade workers in the design of granaries and grain preservation. Occasionally, they are required by the state to train personnel in other state granaries.

Some of the older and experienced workers at the Shanghai Mill stay, in part, beyond the retirement age of 60 to train new workers. Last year, the mill established a workers' university (July 21) to train young employees. The training on the job lasts 2 years. Twenty full-time students were enrolled. They participated in design and conversion of half the mill for pneumatic conveying. Personnel in the bakery at the I-Li factory received some training at the university; most of the training was on the job.

7

WHEAT PRODUCTION AND DISTRIBUTION

The major increase in wheat production since 1949 has come largely from increased wheat yields per sown hectare. If yield in 1949 is multiplied by the wheat-sown area for 1957, the percentage output increase from sown area growth accounts for only one third of the actual wheat output expansion that took place during the period. Sometime after 1957, probably after 1966, the wheat-sown area began to decline so that the output increase must be accounted for by yield increase. However, China may have now reached the limit, or nearly so, in multiple cropping for certain sections of the country. In these areas at least, land is again somewhat "fixed," and future production increases must depend increasingly on improved technology designed to get more wheat from each crop. It would seem that improved and expanded technology in fertilizer and water management, together with continued control of diseases and continued search for ever better-adapted, higher-yielding varieties, offers the best chance for increasing per-unit yields in China. The potential is there, and the Chinese people seem determined to reach that potential.

PRODUCTION, SOWN AREA, AND YIELD

As the second most important food grain in China, wheat has recently occupied at least one fifth of the sown area devoted to food grain production. In Table 2 we observe that the wheat-sown area gradually increased during the 1950's and then declined during the 1960's. By the late 1960's new high-yielding varieties, disease control, improved irrigation, and increased application of organic fertilizer had increased yield on the better soils. Further, wheat production during this same period more than kept pace with the growth of total food grain. The share of wheat of food grain output increased to 14 percent in 1972 compared to only 13 percent in 1955-57 and 12 percent in 1966.

Certainly, the average wheat yield of 1972 appears plausible. Wheat yield data obtained from visits to 10 communes and production brigades in 1976 were very high and averaged around 5,000 kg/ha. These sample areas undoubtedly represent areas of best farming practices. Their yields were roughly 4 times the average wheat yield in 1972.

TABLE 2 Wheat Production and Its Relationship to Food Grain, 1949-1957 and 1972

Item	1949-1951[a]	1952-1954[a]	1955-1957[a]	1966[b]	1972[b]
Wheat production, metric tons	15,183,000	19,916,000	23,800,000	25,179,000	35,880,000
Wheat-cultivated area, ha	22,457,000	25,794,000	27,184,000		26,000,000
Wheat yield, kg/ha	652	770	876		1,380
Wheat output as a percentage of food grain output, %	12.38	12.67	13.15	12.30	14.20
Food grain production, metric tons	122,616,000	157,250,000	180,933,000	204,265,000	252,676,000
Wheat area as a percentage of food grain-sown area, %	21	23	22		20
Food grain-sown area, ha	104,470,000	114,305,500	121,192,000		130,000,000

[a]SOURCE: Nai-ruenn Chen, ed., *Chinese Economic Statistics: A Handbook for China* (Chicago: Aldine Publishing Co., 1967), pp. 338-39 for output data; p. 318 for yield data; pp. 286-87 for sown area data.

[b]SOURCE: Shan-tung sheng lai-yang nung-yeh hsüeh-hsiao, *Hsiao-mai* (Wheat) (Peking: Hsin-hua shu-tien, 1975), p. 1. (Wheat output in 1972 was supposed to be 1.6 times the wheat output of 1949. Wheat production in 1972 was then supposed to be one seventh that of food grain production in 1972. The wheat-sown area was reported to be about 390,000,000 mu or 26 million ha.)

GEOGRAPHIC DISTRIBUTION OF SPRING AND WINTER WHEAT

Wheat, or *hsiao-mai*, occupies roughly 20 percent of the land used to grow food grain and follows rice as the second most important crop. Although both spring and winter wheat can be found, the winter type is more widespread, being grown primarily between the latitudes of 20° and 40°. There are distinctive regional variations in the kind and amounts of wheat grown; Table 3 and Figure 18 present some of these general characteristics. Note that nearly two thirds of the total wheat production and cultivated area is confined to the area regarded as the North China Plain region. Nearly 30 percent of the wheat-sown area and production is found in the central and southern provinces. The remaining parts of the country specialize in spring wheat, and while the expansion of such cultivation has been greatly encouraged in the past quarter century, the bulk of wheat produced is located in provinces coterminous with the major farming regions, which are also the most densely populated areas of the country.

THE ORGANIZATION OF WHEAT PRODUCTION

There are 1-, 3-, and 5-year production plans, which are first formulated for the country and then for the region on a provincial basis. The basic coordinating unit is the county (*hsien*), which after receiving a planned target for wheat and other crops then attempts to assess local conditions to determine if the targets can be achieved and how the

TABLE 3 Wheat-Sown Area and Production in Ten Wheat-Growing Areas (1950's)

Map Identification	Wheat-Sown Area, % of Area's Cultivated Land	Wheat Cultivation Area, % of China's Wheat-Cultivated Area	Regional Share of China's Wheat Production
1	24.5	12.5	10.1
2	32.4	46.5	45.1
3	35.3	1.1	1.7
4	13.5	22.7	21.7
5	10.2	4.2	6.5
6	4.4	3.0	1.7
7	5.8	3.9	4.2
8	15.4	4.6	6.6
9	46.3	1.4	2.0
10	28.7	1.1	1.4

SOURCE: Chin Shan-yu, comp., *Chung-kuo hsiao-mai tsai-p'ei hsu* (Studies of Wheat Cultivation in China) (Peking, 1961), vol. 1, pp. 27-43.

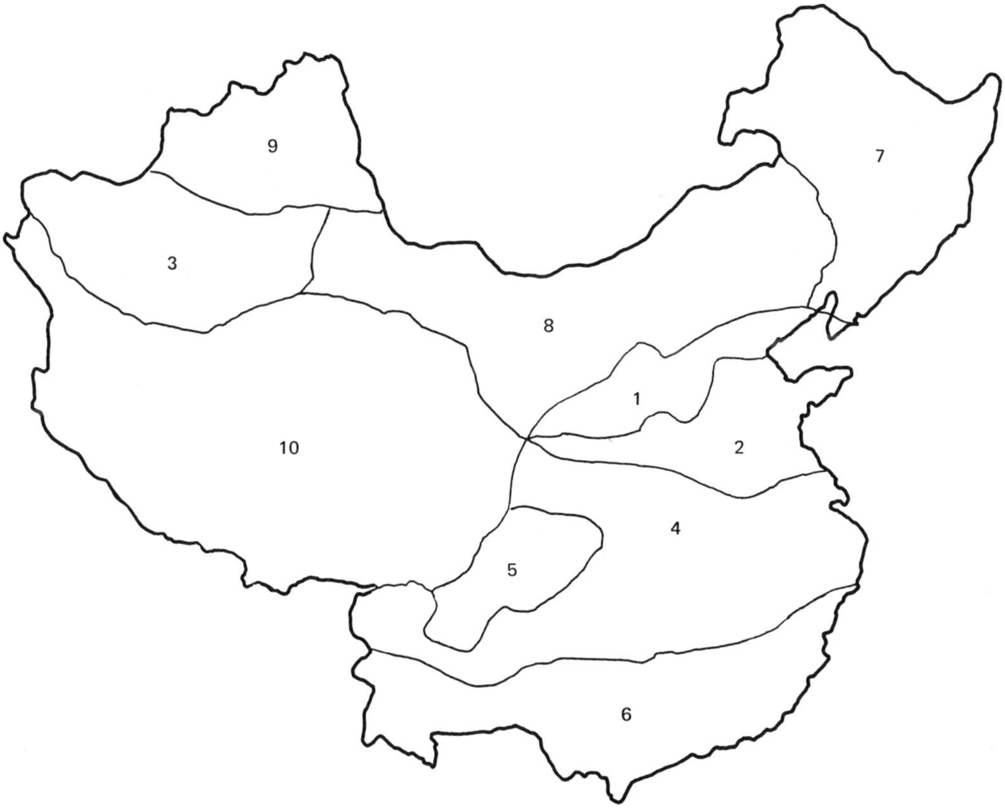

1 Late-ripening winter wheat (Peking region)
2 North China Plain winter wheat (Shantung, Hopei, and Honan)
3 Late-ripening winter wheat (southern Sinkiang)
4 Early-ripening winter wheat (Yangtze Valley area)
5 Early-ripening winter wheat (Szechwan)
6 Early-ripening winter wheat (south China)
7 Spring wheat (northeast China)
8 Spring wheat (north China)
9 Spring and winter wheat (northern Sinkiang)
10 High-plateau spring wheat (Tsinghai)

FIGURE 18 Wheat-sown areas of China.

targets will have to be modified. The county does this by coordinating the activities of the second, basic planning unit, the commune (*kung-she*), which is a territorial unit roughly equivalent to the old standard marketing town and its satellite villages.

Several noteworthy features emerge from the advanced farming communes visited by members of the Wheat Studies Delegation (see Appendix D). First, except for northeast China, the population density of these communes is very great. Second, the number of persons per household is slightly below the prewar rural average household size as reported in the studies by J. L. Buck and other rural investigators. Third, the area of cultivated land is very small for such a huge population size. Within the past decade this land area has not increased; in some instances, it even declined when land was withdrawn from farm production because of the construction of new structures or roads. If these communes represent the major farming areas for population pressure on limited farming land, the problems of planning sufficient food supply indeed become complex and difficult.

The organization and method of planning wheat production are as follows: At the end of the harvest, commune leaders meet with county officials to discuss the new wheat target figures. The commune leaders then meet with production brigade and team leaders to determine how these targets will be met. By holding formal conferences to assess the available input resources for wheat to produce the desired targets, the commune leaders achieve an informal basis by which to discuss with county officials the prospects for meeting planned wheat output targets. Numerous conferences involving administrators, technicians, and accountants are convened to work out the details as to how much land and other resources will be devoted to wheat production to produce the stipulated output target. The agreed upon output figure for each commune included the contribution both brigades and teams will make. In addition, this planned target figure takes into account, in a very general way, the amount the commune will sell to the state and the remainder as income and food ration. After the next wheat harvest the sown areas, yield, output, sales to the state, and retained amount along with evidence of the allocation and amounts of inputs for wheat production are recorded and the information on harvest performance is then made available to the county administration for reporting to the state. Figure 19 shows this formal, bureaucratic system at the county level for Chiang Ning Commune (near Nanking) by which the planning of wheat production is carried out.

It should be further mentioned with reference to Figure 19 that the county plans and coordinates the building of wells in communes and the production and distribution of chemical fertilizer, as well as other farm inputs. It also carries out extensive seed-breeding work. The numerous, formal conferences of commune leaders with the county officials serve to coordinate, implement, and record the results of the various agricultural plans. New committees and job assignments have led to a certain amount of bureaucratization at the rural community level.

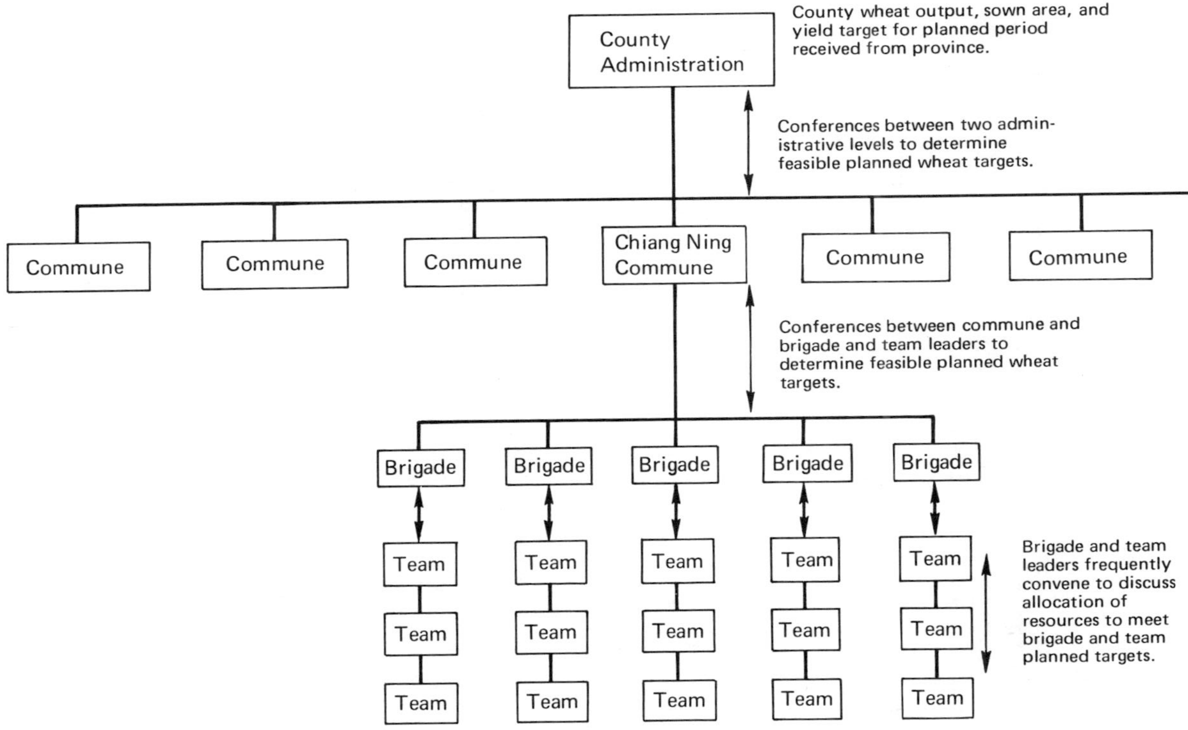

SOURCE: Field notes from visit to Chiang Ning People's Commune, May 26, 1976.

FIGURE 19 Formalized planning of wheat production in Chiang Ning County, Kiangsu, mid-1976.

WHEAT DISTRIBUTION

Harvested wheat is collected and used for different purposes. First, a fixed amount is sold to the state at the stipulated state price. Communes have the option to sell above the fixed quota if they desire more money income. Second, each household is supplied a specified grain ration, comprised in part of wheat, which it will take to the team, brigade, or commune flour mill to be processed. Third, the commune stores some grain as a provisional stock for contingency situations to be distributed as food ration or to be sold for money income for usages dictated by the commune.

In the communes visited by the Wheat Studies Delegation we observed that only in communes that specialize heavily in wheat can a significant share be sold to the state. Wheat, like rice, is a food grain, and it was difficult to learn the precise share of wheat sold, since communes generally reckon in units of food grain rather than a specific food crop. There was much similarity in wheat prices sold to the state. Even on a regional basis the price variation was small. Furthermore, wheat prices have been remarkably stable and have not changed in the last 4 or 5 years. The amount of wheat available on a grain ration basis appears to be sufficient for consumption purposes. If these conditions of wheat and/or food grain availability are representative of the rest of China, these figures would indicate that food supply is quite adequate. In fact, we were informed that available food grain per capita in communes of the Kirin Province in 1975 was 600 kg. Allowing that 100 kg were sold, and another 200 to 250 kg would be used for seed and feed, the residual of around 250 to 300 kg of food grain per capita for this northeastern province compares favorably with the amount of wheat and available food grain for the communes visited by this delegation.

We were told by agricultural economists of the Northwest Agricultural College at Wukung (Shensi) that one third of commune food grain production went to the state, another third was used for seed and forage, and the remaining third was consumed by commune members. It is improbable that all communes sold as much as a third of the grain produced, and a share more nearly one fifth or one fourth seems more appropriate.

Wheat sold to the state enters the urban storage, milling, and distribution channels. Communes are responsible for maintaining quality control of the grain shipped to the state, and they also pay for shipping charges and handle the transportation of the grain to state storage centers or granaries. These granaries are under the management of the State Grain Bureau. Each city has a grain bureau, which regulates the activities of granaries, flour mills, and retail outlets. The system by which the city of Sian with 2.5 million population received and stored its grain in 1976 is shown in Figure 20. The food grain bureau controls prices, storage, grain transfers, milling standards, and grain availability.

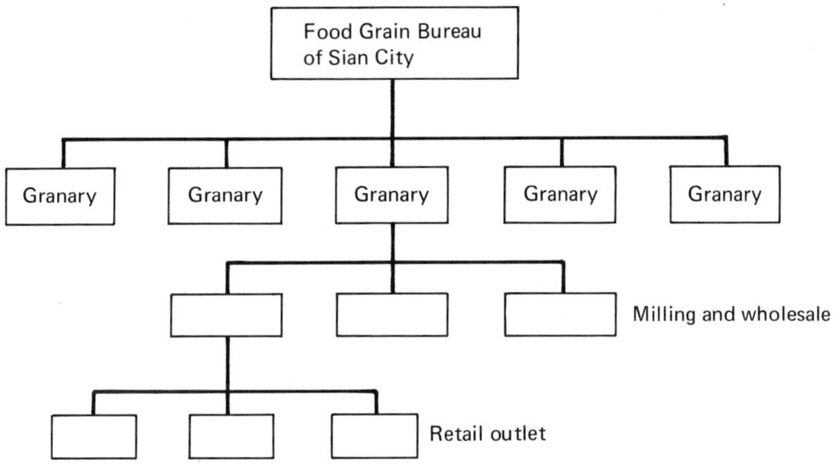

FIGURE 20 Urban food grain storage and distribution, Sian, 1976.

The Campaign to Learn from Tachai

For more than 10 years the party has promoted the new agricultural policy of learning from the experiences of the Tachai production brigade in Shansi Province. In some years this policy has been promoted vigorously, but in other years less so. During the late spring and early summer of 1976 the promotion was pushed at all levels in the countryside from the agricultural research institutes to the smallest production team. Essays in university journals, statements in the press, and slogans on walls and fences attested to the importance that the party assigned to this campaign. The implications of this policy upon wheat production and agricultural organization may be several.

Steps are under way to shift the accounting unit from the production team to the production brigade. The brigade will then control the allocation of resources to teams and the distribution of income to team members. Some production brigades near Sian and Shihchiachuang visited by the Wheat Studies Delegation already have implemented this new accounting procedure. This is a change that may be resisted in many areas because of the unequal distribution of resources, especially labor, between production teams. When the vice-chairman of the Shuang Chiao People's Commune outside of Peking was asked if his commune intended to make such a shift, he replied that they had no plan in the immediate future to do so. His explanation was that the existing differences between production teams were large and, until these were removed, the shifting of accounting to the brigade level would be difficult.

The campaign also seeks to rebuild every rural community in the country. This will be done by creating a special work team in each production brigade and by using part of the brigade accumulation fund to construct new houses for each family. These same teams also will build schools, recreation halls, factories, and office buildings in the brigade. These self-contained communities with their service facilities constitute a kind of "urbanization of the countryside" in that the amenities of the city will have been created within densely populated rural communities. The party hopes by this step to improve farmer incentives to work harder and restrain other forms of consumption spending. Further, this massive rebuilding program presumably will be done in a spirit of self-reliance without state assistance.

The campaign also implies a major attempt to mechanize farm production with machines such as ploughs, rice transplanters, harvesters, and tractors. The laborers freed by such machines will be utilized for the rebuilding program cited above and will enjoy more leisure time as well.

There are significant implications for rural people from the learn from Tachai campaign which suggest more bureaucratization of life and work in the countryside. The new self-contained communities will be carefully regulated and planned by organizations within and above the commune administration. Individuals and families are not allowed to accumulate production assets.

FACTORS ACCOUNTING FOR WHEAT PRODUCTION DEVELOPMENT SINCE 1949

Although the wheat-sown area rapidly increased during the decade of the 1950's, it stabilized at some point in the 1960's and then declined so that by the early 1970's it approximated that of the mid-1950's. The wheat yield of the current sown area is about twice that of the 1949-51 period.

The absence of detailed, accurate agricultural statistics related to wheat production makes it impossible to use production function analysis to identify and assess the contribution of each factor to increased wheat output. It is possible to classify and discuss these factors in a general way by distinguishing between the two broad categories of conventional and modern farming inputs. Conventional inputs are defined as land, labor, and capital produced within the farming sector by traditional methods. Modern inputs are new forms of capital that originate outside the farming sector and are produced by modern technology. Some modern farm capital is now gradually being selected and produced within Chinese agriculture and we anticipate that this will continue.

Conventional Farming Inputs

Although cultivated land increased slightly during the 1950's, by the mid-1960's the Chinese were concentrating their efforts and resources on increasing the sown area of their existing cultivated land rather than extending that land. The present campaign to learn from Tachai aims at reclaiming marginal lands for farming. The major thrust of that policy is to improve existing land by different organizing efforts and by a new work spirit. In the communes visited by the Wheat Studies

Delegation we learned that some sections of fields had formerly been hills or gullies. Labor teams had leveled or supplied earth as fill-in in order to maximize the area of land that could be farmed. At the same time, labor teams increased irrigation facilities and improved drainage. These actions enabled communes to retain the same amount of cultivated land as in the mid-1960's, even though some farmland had been withdrawn from production because of road and building construction.

When private farming was abolished in the mid-1950's, farm labor was mobilized for nonfarming activities during the winter periods. The immediate effect was to increase the days that an able-bodied farm laborer worked to 300 per year from less than 300, as was customary during slack periods under family farming. Another major development of equal importance has been the formation of teams of women who engage in farm work to free teams of men for capital construction projects. Older people also have been mobilized within the household to care for children so that women can and do work alongside their men in the fields. During the winter months and between harvest and next planting, teams of workers build terraces, dig wells, construct houses, level fields, etc. As a result, the new team-farming system has mobilized more labor than that of the family-farming system before it. Work and leisure time now are more regulated and controlled by group organizations within the commune.

The consequence of such labor mobilization efforts has been to gradually increase the traditional forms of farm capital such as field wells, drainage ditches, supply of organic fertilizer, farm structures, and farm tools. Examples of this development were in ready abundance on the advanced communes visited by the Wheat Studies Delegation. We were informed that the quantity of organic fertilizers applied to the land had been gradually increased since the mid-1960's. New wells also had been constructed, and many new farm structures such as grain storage buildings, small flour mills, and livestock dwellings for hogs, chickens, and ducks had been built.

Modern Farm Capital

There are two basic types of modern farm capital that affect crop production in China. The first includes such inputs as new seed varieties, chemical fertilizers, and new knowledge about soil preparation, planting of seeds, and crop management. These inputs vitally affect the quality of crops and their yields. The second deals with the human and animal energy required to farm the land. Here the use of machines to conserve human and animal energy so as to direct it to other uses becomes important. New machines that provide more power for transportation and that save labor and animal energy at various stages of farming include tractors, large ploughs, rice transplanters, harvesting machines, etc. These types of modern capital make it possible for less labor and less land to produce more.

Evidence gathered by the Wheat Studies Delegation strongly suggests that for wheat production (and this probably applies, as well, to rice, fiber crops like cotton, and some oil seed-bearing crops like soybeans), there has occurred in China over the past quarter of a century significant advances in development and use of new seed varieties but less progress in modern fertilizer technology. The former, in part, originated from the foundations established during the 1920's and 1930's and successive developments thereafter during the 1950's and 1960's. The large number of new wheat varieties developed, tested, multiplied, and introduced for commercial production since the 1950's attests to the activities of the research institutes and the capabilities of agricultural colleges and extension services of the county-commune rural organizations.

On the other hand, the reliance on green manure crops in crop rotation systems, the heavy use of organic fertilizers, and the application of only small quantities of chemical fertilizer with low nitrogen content indicate that the chemical fertilizer revolution has not yet occurred. If these conditions that characterize the most advanced farming communes near the largest metropolitan centers apply elsewhere in the country, especially to the poorer, more backward farming districts, then an increased supply of high-quality chemical fertilizer in the future would have a large impact on crop yields and output and thereby dramatically affect China's food supply and foreign trade prospects.

In the second area of modern farm capital there is convincing evidence that a gradual revolution of farm machinery is beginning to occur. In some advanced communes there are workshops and small factories that now produce a variety of modern farm tools and machines both for use on the commune or to be sold to other communes. Large tractors, however, still must be purchased from modern industry located in the large cities.

The information that we acquired about large farm machinery was given only upon request. The very advanced commune like the Sino-Albanian Friendship Commune already is using machines for nearly one third of its field operations, but such cases are still extremely rare. A more reasonable conjecture would be that among the most advanced communes, and it is difficult to ascertain how many communes in the country ought to be included in this criterion, perhaps only about 15 or 20 percent of the land and field operations are mechanized. But a beginning has been made, and the Chinese intend to increase the supply of farm machinery to reduce the back-breaking toil of the farming team.

THE ORGANIZATION OF TEAM FARMING

Between 1958 and 1961 the Chinese Communist Party created the commune administrative unit, which covered an area equivalent in size to the traditional standard market town and its satellite villages. This new unit now coordinates, regulates, and controls the many hundreds of villages and towns within its boundaries. The villages within each commune were renamed brigades, and many were combined into large rural communities. Each brigade organized production teams, which were groups of households clustered in natural neighborhoods within each village. By 1957, village households had been combined into farming teams, each with a leader, an accountant, and a secretary. From 1961 to the present, farming within the county and commune has been organized on a team basis within brigades, which in turn are accountable to the commune and county administrative units. While the commune serves in many ways as the lowest level of the state administration system, it also has some independence from the county. Its staff, paid in part from commune income and selected by commune members, still must be approved by the county.

Farming teams are coordinated according to annual plans and regulated by procedures formalized within the bureaucratic administration of the commune and county. At the present time the party is urging commune and brigade leaders to move the accounting unit from the team to the brigade level. For example, at the Sino-Albanian Friendship Commune, eight of the 28 brigades in the commune have shifted the accounting of team resources and work points from the team to the brigade. It probably will be at least another decade, or longer, before the brigade unit will fully regulate and control the income and distribution of all of its households.

There are advantages in having the county office work through the commune system to control the activities of rural communities. These advantages work toward the benefit of the state and, in some instances, of commune households as well. The state can plan the development of well-defined economic regions by having counties coordinate commune capital projects or combine their efforts for large-scale water control projects like that of the Hai River project. During the late 1960's and early 1970's, commune labor in many counties was mobilized throughout central Hopei to build 34 large canals and construct 4,300 km of flood control dykes for this project. The external economics of such water control projects would be very large for the many communes severely threatened by floods each year. The state also benefits by obtaining higher taxes and grain procurements from that production saved from flood and drought.

The county and commune organizations now operate an extension system for selecting, testing, and increasing new wheat varieties and other crop seeds for use in commune fields. The system serves as a conduit for channeling the research results of higher educational and advanced research units to the farming areas. It allows communes to utilize varieties best suited to local conditions and derive early benefits from them. At the same time, interaction between institutes-colleges and communes is increased. The former send technical personnel to the communes to lecture in the short-term agricultural colleges that are now mushrooming and advise commune agricultural technicians. The latter send their experts to research centers to learn or work on specific problems.

Team-Farming Incentives

The present team-farming system provides both incentives and discouragement to household members now farming the land. Under the old family-farming system, families could freely accumulate property and decide the use of household resources. Many households often became destitute when engulfed by natural calamities because either the efforts of the state or kinship ties proved to be too late or too little to assist them.

Under the team-farming system, no household can become destitute because some redistribution from commune income will always take place to assist needy families. On the other hand, many families may be discouraged from working harder or using their managerial skills because property, with the exception of one's dwelling, cannot be accumulated as in former times and the rewards may not always be self-evident. It is difficult, however, to assess the impact of improved incentives or prevailing constraints that discourage families from participating more or less vigorously in the team-farming system. Some examples can be cited within the commune organization that may account to some degree for the party having been able to mobilize the rural people to work diligently throughout most of the year. We list four major considerations.

The supply of food grain available for commune members on a per capita basis has gradually but steadily increased and has become more stable each year. An adequate supply of food grain each year reduces family fears and uncertainty about whether or not enough food will be available at the end of the farming system.

Household income and savings have very gradually risen during the last decade. Farm prices and input costs have remained relatively stable in the short run. As production has gradually increased, communes have been able to allocate more income to their accumulation fund as well as to distribute more to households. Although variation in per capita amounts of distributed brigade and commune income has existed in regions, per capita income has slowly risen. Further, the share of income for the accumulation fund has also varied greatly. A probable reason that the Ta Yu Shih production brigade had a high accumulation share is that this brigade has begun to rebuild every residential home. A special construction team with equipment has been established for this purpose.

The land tax and crop procurement quotas have had the effect of lessening their burden upon commune members in that as production and income rise, the burden declines. The land tax, fixed and assessed in the early years of the regime, declines in relation to other tax burdens each year. Crop procurement quotas also remain fixed over long periods so that once these are met, communes can freely distribute or sell the remainder. The share of total income distributed to brigade or commune members differs greatly, ranging from 40 to 60 percent. A reason for high production costs such as those at the Sino-Albanian Friendship Commune is the great expense of maintaining a large supply of farm machinery.

Communes have made a special effort to improve public health and medical services and educate their members. Advanced communes now have clinics even at the production team and brigade levels. There are primary, even secondary, schools and some agricultural technical schools in these same communes. The increased supply of these services has contributed toward improved health of males and females of all ages.

Finally, there is a movement under way now to rebuild every residential dwelling in the countryside. It is not clear in every instance whether rural people will always own their dwellings. For example, in the Ta Yu Shih production brigade in Kirin, brigade members pay an annual fee to the brigade to live in their dwellings, but the brigade in essence retains ownership rights over all homes. Visits to Ma Lu and Chiang Ning communes in Kiangsu revealed that households use their savings and borrow from the commune to rebuild their homes and thereby retain ownership over their dwellings. Whether or not ownership of residential housing will be widened or narrowed as the trend of rebuilding homes continues is not clear at this time, but the fact remains that rural houses are being rebuilt all over China. This development raises the material living standards of the people and provides them with tangible evidence that progress, while slow, is taking place and must be attributed to the commune system of farming and organized rural life.

The above incentive factors in part explain why rural people participate in team farming and are willing to work most of the year (300 or more workdays per year). What cannot be ascertained from our brief travel in the countryside of China is the spirit and ideology of the rural peoples and their likely behavior if they were given the opportunity to participate in on-going commune organizations and permitted to expand their private plots, farm on a household or kinship basis, and market their produce through state channels.

Since the early 1950's the supply of conventional farm inputs in China has been gradually increased. In particular, the supply of farm labor has been mobilized and utilized more intensively than under the former system of family farming. Farming teams, as a result, have increased the supply of traditional farm capital, an increase which, in turn, has contributed enormously toward improving the quality of the land. State organizations have continued to make available, with slight interruption only between 1965 and 1971, new, early maturing crop varieties that have raised the sown area index and changed the crop rotation systems during the late 1960's. There are still large differences in crop yields within communes and provinces, an indication that much potential still exists, although with considerable cost, to raise crop yields at the lower end of the scale. For example, on the Sino-Albanian Friendship Commune it was reported that some land still produced only 1,125 kg/ha of wheat, whereas the average wheat yield was 3,500 kg/ha. Within Hopei Province the high, stable crop yield communes visited by the Wheat Studies Delegation averaged around 5,000 to 6,000 kg/ha, but in 1975 the average yield in the province was reported to be only 1,646 kg/ha as compared to a yield of 713 kg/ha in 1957. The potential to raise yields can and will be realized when the supply of chemical fertilizer of higher quality and nitrogen content becomes available to the low crop yield areas. On the other hand, it may be much more difficult to increase the already very high yields in advanced communes without more research on the complex variables affecting productivity.

WHEAT CONSUMPTION AND WHEAT IMPORTS

In former times much of the wheat produced moved to the large cities for milling and distribution to city dwellers. Farmers cultivated wheat mainly as a cash crop to be marketed. In the south, wheat moved by waterway to Hankow, Nanking, cities on the Soochow plain, and

Shanghai, whereas in the north, wheat moved by railway to Taiyuan, Paoting, Kaifeng, Peking, Tientsin, Tsinan, and Tsingtao. After World War I, the import of wheat increased greatly, mainly because of the disruption to transportation and marketing. After 1949 the restoration of peace and order and the recovery of agriculture to the prewar level made it possible to eliminate wheat imports for a decade. After 1960 China again began to import wheat in the amount of 3 to 6 million tons per year. These imports continued into the 1970's although at a slightly reduced volume. They are expected to continue into the near future.

As was mentioned already, the present sown wheat area is smaller than that of the late 1950's, but wheat production and yields have risen during the period since the 1950's. Meanwhile, population also has increased, and while migration from the countryside to the cities has been halted, urban population continues to expand but at a rate now below that of population growth in the countryside. What explains the fact that China has imported wheat for 15 years? To answer this question, another must be posed. Where is the bulk of imported wheat consumed? Our answer is purely speculative, but on the basis of a visit to one flour mill in Shanghai and general observations of the current state of transportation, we would suggest that imported wheat is consumed mainly in the large cities of Peking, Tientsin, and Shanghai.

The Shanghai flour mill visited by members of the Wheat Studies Delegation sheds some light on the present wheat processing and distribution system in the large cities of China. We do not know the exact number of flour mills in Shanghai, but we will assume that their operations are similar to the mill we visited and that variation in wheat capacity and flour production is not large. The mill we visited had 1,600 workers, having been built in 1936. In 1958 it produced only 1,800 tons of flour per day, but in 1975 it produced 2,500 tons of flour per day. If we allow for an extraction rate of 80 percent of flour and 20 percent of bran, from the annual production of flour of 912,500 tons, we estimate the required wheat for processing to be 1,140,625 tons. The storage capacity of the mill was 24,000 metric tons, which means that at least 38 times a year or 3 times a month, wheat must be supplied to the mill for milling. The mill worked three shifts a day, a worker laboring 8 hours per shift. Mill officials reported that this mill obtained about 50 percent of its wheat from abroad. This would mean that the mill used about 456,250 tons of imported wheat. Assuming that there are at least five flour mills of similar size in Shanghai, perhaps as much as 2 million tons of flour could be produced in this city alone, which would range from one third to one half of total imports. The remaining imported wheat would have been processed and consumed as flour and noodles in Tientsin and Peking.

Inasmuch as these large cities possess adequate wheat storage capacity, the question then becomes one of the following. Is wheat imported and consumed in large cities because the domestic supply of wheat is inelastic and/or because high transportation costs prevent wheat from being shipped great distances to centers of large population demand? The figures obtained for food grain and wheat availability on a per capita basis for the advanced communes visited by the Wheat Studies Delegation clearly indicate that rural communes around large cities supply enough for their annual needs. Further specialization in wheat might be possible in certain areas, even at the expense of other crops, but perhaps the cost of transport plus the opportunity costs of growing wheat are so great as to exceed the cost of importing wheat from foreign sources. If this argument has any validity, it would mean that wheat imports would only be eliminated when (1) present wheat production rises much more rapidly than does domestic demand, (2) transport costs decrease, or (3) production of other food grain crops rises more rapidly than does demand. Inasmuch as the large cities in the interior of China are supplied by an adequate amount of wheat from within each province, and this seems true for Shihchiachuang, Sian, and Nanking, which we visited, then the decision to import some wheat for the extremely large cities of Tientsin, Peking, and Shanghai is probably due to the high costs of transport and shifting land from other crops to wheat.

8

FARM LABOR AND LIVING IN CHINA

Wheat growing in particular and farming in general must enable people to make a living--like any other productive activity. There must be adequate opportunities for work, which must yield appropriate returns to efforts, so that satisfactory levels of living can be achieved by those engaging in farming. Otherwise, these people may be expected to abandon their efforts, shift to other work, flee the countryside in search of better opportunities, and possibly pose problems of absorption and social integration elsewhere.

By all indications, old China experienced such problems, just as many underdeveloped countries do today. In contrast, new China, in the opinion of its leaders, has overcome these difficulties, not only by imposing restrictions on the farmers' residential and occupational mobility. Characteristic of the whole approach to rural development is the negation of "zero marginal productivity" arguments in their usual form. According to the Chinese, farm labor productivity and farm income can be increased substantially by the exertion of more effort, even in the absence of modern means and methods of production. Such a view does not imply, and the Chinese have not inferred from it, that the modernization of agriculture should be forgone or greatly delayed. But it does mean that the transition to modernity may follow paths that until now have not been considered traversable by the rest of the world.

The Chinese claim that they follow this approach to agricultural development everywhere, in all circumstances. But they practice what they preach with the greatest determination in numerous model communes, some of which were shown to the Wheat Studies Delegation. The accomplishments of these select units are widely publicized, as proof of the "correctness" of their strategy and as exemplary achievements that everyone else is encouraged to emulate. In contrast, the less impressive performance of the farm sector as a whole is not so prominently reported and never in comparable detail. Needless to say, such selective publicity makes it difficult to place our firsthand observations and reports into proper perspective.

To form an opinion about the potential effectiveness of the approach, data on the present accomplishments of the inspected model communes will be compared not with what little was made known to us about their past performance, but with information from John Lossing Buck's comprehensive sample surveys, *Chinese Farm Economy 1921-1925* and *Land Utilization in China 1929-1933*, which describe the state of farming prior to Liberation in the same or similar terms.

Of course, this conceptual correspondence does not solve all problems of comparison. The data that the Wheat Studies Delegation collected pertain to a small number of model communes in a few locations, whereas the data that Buck compiled refer to a large sample of perhaps not so unusual farm households in many locations. Only in two instances do Buck's data and our data refer to the same *hsien* (administrative district), though not necessarily to the same location within it. And in all cases our data are much more "casually empiricist" than are Buck's meticulous statistics. As a consequence, one may perhaps wonder about the significance of the exercise.

Fortunately, the relevant contrasts between the sample of old farm practices and the models of the new agricultural methods are so stark that the advantages of the latter appear not in doubt. However, it is also evident that the accomplishments of these models are not representative. Therefore, besides recognizing the worth of the new approach, one must in addition explain its limited effectiveness at present. Unfortunately, the Wheat Studies Delegation received very little information in response to this question. Indeed, much of what may be said in answer to it depends really on sources with which we became familiar outside rather than inside China.

POPULATION AND LABOR FORCE

Occupational Structure

Rural people's communes as well as agricultural production brigades include not only farm but also nonfarm households. Moreover, the latter are particularly numerous in the outskirts of cities, where a substantial share of the communal labor force has been transferred to work in industry either within the commune or in a state enterprise on the outside. Since several of the communes that we saw were located in such environs, it is not surprising to find that the manpower shares of their farm employment are substantially lower than the 80 percent share of farm work that Buck derived for his sample of farm households. What is impressive, however, is the change in the composition of other rural activities and occupations. Traditional forms of commerce and handicrafts appear to have diminished greatly everywhere, while industry proper and modern services, notably in health and education, have come to the fore, at least in the collective sector of the economy of the commune or brigade.

Labor Force Share

Buck's findings on average household sizes in the regions where the visited communes are located exceed those that the Wheat Studies Delegation obtained, and the difference between his 6.0-6.7 persons and our 4.0-5.1 persons per household cannot be explained simply with references to changes in the occupational structure. However, the number of his "man equivalents" per household differs similarly from the number of the communal "labor powers" per household, and the shares of the labor force in the population fall within similar ranges near 40 percent. Slightly lower ratios in the communes may reflect the different measurement of manpower units as well as high population growth rates since the 1950's, which made for larger shares of young dependents.

Yet there is one exception. The 26.7 percent share of labor power in the population of Nan Wei Tzu People's Commune cannot be explained in this manner but seems to be attributable instead to an unusually low rate of female labor force participation. Elsewhere, where the share of labor power in the communal population was at least normal, it was said that 96 percent of all women under 45 years of age participate in production, that 55 percent of the communal labor force are female (while many males work outside the commune), etc. More detailed data on the sex composition of the population and labor force were not obtained.

In general, Buck's observation that "virtually every able-bodied person seven or more years of age worked" is still true, and probably even more so than when it was first made. Of course, those aged 7-15 are found in school today during 9 months of the year, and their participation in production is limited at that time. School closes for the summer months in order to release the students for work in the fields.

Work Efforts

The statement that almost everybody worked in old China does not indicate the extent of such efforts. Other data suggest that it was quite limited. Able-bodied men of ages 15-60, who worked about 10 months per year full-time or part-time, performed close to 80 percent of all the farm work according to Buck. By implication, each man equivalent contributed hardly more than 6 months of work per year on the average. Moreover, when the days of work were measured more precisely as man work units (i.e., 10-hour days), the average effort was even less, viz., 120 days per man equivalent per annum.

In new China, by contrast, we learned that commonly, people, i.e., both men and women, now work at least 300 and often 310 or 320 days per year, no matter what the occupation. Only the children, because of their school attendance, continue to contribute less, probably about 100 days as before. The average number of workdays per member of the communal labor force has thus at least doubled and perhaps increased by more, while the share of labor force members in the communal population may not have risen greatly. The indications are that this increase occurred gradually during the course of collectivization and communization, notably during the Great Leap Forward, when women were encouraged to leave domestic work for the fields.

POPULATION-LAND RATIO

The increase in effort evolved while the amount of land per member of the population (as well as per member of the labor force) declined. During 1929-33, Buck's farms in the regions where the visited communes are located cultivated about one-half acre (= 1/5 hectare = 3 mu) per capita. Only in the spring wheat region was the average crop area larger at three-fourths acre (= 3/10 ha = 4.5 mu per person). In contrast, the communes that we visited in 1976 listed average areas of 1.0 to 1.8 mu or 0.07 to 0.12 ha per capita, with two exceptions. Nan Wei Tzu People's Commune in the spring wheat region has almost 3 mu or 1/5 ha of cultivated land per person, whereas Huai Ti brigade in the outskirts of Shihchiachuang has been reduced to 2/3 mu or 0.04 ha per capita of its population.

Additions to Land

Apart from sampling problems, the contrast reflects a substantial increase in population relative to cultivated land. The prospects for additions to the latter were already dim in the 1930's, when Buck guessed that the removal of graves and boundaries plus defragmentation, reclamation, and the consolidation of farmsteads might increase the area of cultivated land by 10 percent. Of course, all of these changes were made during the years of collectivization and communization. At the same time, however, urban growth and industrialization diverted land from agricultural to nonagricultural uses. As a consequence, the net additions to farmland appear to have been relatively small.

Population Growth

In any case, the land additions were small in comparison with the additions to the population who had to live on as well as off the land. Buck found for 1929-33 a crude birth rate of 3.8 percent and a crude death rate of 2.7 percent. The restoration of peace and order plus the establishment of a remarkable public health system--likewise during the course of collectivization and communization--reduced the death rate rapidly but did little at first to change the birth rate. As a result, the rate of population growth rose--to 2 percent or more per annum. Population developments have not been reported since the 1950's. Data on student enrollments imply that the birth rates in the visited communes averaged 2.5-3.0 percent during the 1960's, with one exception below 2.0 percent. Most recently, however, population planning seems to have become extremely effective everywhere. For 1975, the visited communes reported extremely low birth rates of 0.5-1.5 percent, death rates of 0.2-0.8 percent, and net growth rates of 0.2-1.0 percent. There thus may be an end in sight to the rise of the population-land ratio.

LABOR UTILIZATION

A similar growth of the rural labor force and the increase in the number of workdays per rural labor force member combined to enlarge the "supply" (availability) of farm labor greatly--both *in toto* and per unit of land. Of course, the old methods of production did not "demand" (require) efforts of such magnitude. The implementation of the National Agricultural Development Program, however, depended on them. Four modifications of established farm practices were (or could be) highly labor consumptive, especially in the wheat-producing regions.

Basic Construction of Irrigation Works

According to Buck's observations, perhaps one fourth of China's farmland had been terraced by 1929-33, and 35-40 percent of the entire crop area was being irrigated, without much change during the twentieth century. This national average was approximated in the spring wheat region and surpassed in the Yangtze rice-wheat region, where paddies were common. In the winter wheat regions, where most of the communes visited by the Wheat Studies Delegation are located, the share of irrigated land had risen to merely 10 percent of the crop area by the 1930's.

According to the observations of the Wheat Studies Delegation, much if not most of the arable land of north and central China has been leveled and is being irrigated as of now. The change was brought about within a few years by massive efforts in "basic construction," i.e., by removing hills, filling in gullies, planing land, digging drainage ditches, building irrigation ditches, sinking wells, etc., mostly with traditional tools and little animal or mechanical power.

In all but one of the communes visited, most of the land had been leveled so that it could be irrigated by flooding. During the initial construction phase, which seems to have been completed everywhere in 2 or 3 years, the communes invested heavily in the undertaking, as much as 40-50 percent of their entire workdays. Even now, however, when what remains to be done is maintenance and improvement of the facilities, the communes visited reported allocations of 6-30 percent of their workdays to basic construction, the unweighted average being 20 percent of the days.

Of course, basic construction is undertaken mostly during the off-season, and it is not limited to work in the fields, but includes all forms of earth work, e.g., road and housing construction.

Fertilizer Preparation

The Chinese have been using organic fertilizers in substantial quantities since time immemorial. For 1929-33, Buck determined applications of about 7.5 tons of animal manure and night soil per sown hectare. The average input of all fertilizers, including green manure, bean cake, silt, etc., but excluding inorganic fertilizers, which were practically unknown, exceeded 10.5 tons per sown hectare. Buck noted that the availability of fertilizer varied substantially with the density of the animal population, but he did not calculate the work involved in preparing the fertilizer.

In contrast, the Wheat Studies Delegation heard almost everywhere references to several times this quantity as the apparent standard of fertilizer application under the agricultural development program. The communes visited--and most agricultural research institutes as well--tend to use about 75 tons of organic fertilizer per sown hectare, and in several instances this input is raised to 150 tons. The preparation of such quantities appears to require substantial work efforts, specific references ranging from 1 to 2.5 days per ton and from 75 to 375 days per sown hectare. There was no mention ever of an inability to meet these standards because of inadequate supplies of manure, night soil, and other ingredients. Apparent problems of nutrient shortage are discussed in a different context.

Intensive Cropping

In comparison with most farmers, the Chinese have tended their crops intensively at all times. Buck's data for 1929-33 imply that they spent from 53 to 80 man-days per hectare sown to wheat in the relevant regions, the nationwide average being 65 days of work. The days per hectare of corn were somewhat less while those per hectare of cotton were 2 and those per hectare of rice almost 3 times as numerous. The greater exertions in cotton growing concerned primarily cultivating and harvesting; those in rice growing were made in transplanting and irrigating.

In contrast, the Wheat Studies Delegation was told that the cultivation of "high-yield plots" of wheat required 225-450 days of work per sown hectare in the communes visited, the most frequent reference being to 300 days per hectare. Similar estimates were given for corn, rice, and cotton. This greater and more uniform effort per unit of cropped land appears to be attributable to the fact that all land--not only rice land--is being irrigated, that the same kind of heavy fertilizing is practiced on all crops, and that deep plowing and other forms of more intensive cultivation are used as a rule. The workdays per sown hectare reached 450 in two instances, Ma Chi Chai People's Commune and Huai Ti Brigade, where corn and wheat were being transplanted by hand. And they stayed--or had fallen below--150 in one case, Sino-Albanian Friendship People's Commune, where part of the work had been mechanized.

Multiple Cropping

Growing more than one crop per year on a piece of land was a well-established practice in China, as the 1929-33 ratio of 1.49 for the 22 provinces suggests. However, the practice varied greatly from region to region, given the climate and the state of agricultural science. In the wheat region, it ranged all the way from 1.65 in the Yangtze rice-wheat region to 1.39 in the winter wheat-kaoliang region to 1.18 in the winter wheat-millet region to 1.07 in the spring wheat region. Also, Buck found indications that multiple cropping was becoming more common.

In comparison with these regional averages, the visited communes currently practice multiple cropping to a substantially greater extent and degree. Chiang Ning People's Commune and Ma Lu People's Commune in the Yangtze rice-wheat region now grow two crops of early maturing rice in addition to one crop of winter wheat or barley on much or most of their grainland. The need for seed beds and nongrain land keeps the ratio below 3.0. But it reaches at least 2.2 in Chiang Ning and 2.6 in Ma Lu. Farther north and inland, it is now common to have winter wheat or barley followed by at least one and sometimes two summer crops, the most usual second crops being rice, cotton, and corn. Therefore, on all the communes visited in the other two winter wheat regions--Shuang Chiao, Yung An, Wu Hsing, Huai Ti, Chih Ma, and Sino-Albanian Friendship communes in the winter wheat-kaoliang region as well as Ma Chi Chai Commune in the winter wheat-millet region--the ratio of multiple cropping begins to approach 2.0, the yield data implying a range of from 1.6 to 1.85. In the spring wheat region, of course, multiple cropping continues to be more limited by climatic conditions. But in Nan Wei Tzu People's Commune, too, spring wheat or barley is often followed by corn, unless higher-yielding rice is planted as a single-crop alternative. Furthermore, here as well as in the other regions there occurs a great deal of transplanting, interplanting, and catch cropping of grains and cotton, in an effort to utilize the land more continuously.

Obviously, the more frequent cropping of the land raises the number of workdays per unit of cultivated land accordingly, both so far as the preparation of organic fertilizer and as the cultivation of the crop itself are concerned. On the assumption that these two activities together required 450 days of work per sown hectare on the average, the work per hectare of cultivated land absorbed 990 and 1,170 days in the Yangtze region communes, 720 to 830 days in the other winter wheat region communes, and probably between 450 and 675 days in Nan Wei Tzu People's Commune in the spring wheat region.

Manpower Absorption

Multiplying the average workdays per hectare of cultivated land with similarly rough approximations of the average number of hectares of cultivated land per agricultural "labor power" and adding to this product the share of days of basic construction in 300 workdays per annum yield the following results: The number of days of farm work plus basic construction required per agricultural laborer approximates the maxima of 300-320 days in Chiang Ning, Ma Lu, Sino-Albanian Friendship, and possibly Shuang Chiao People's Communes. This approximately full utilization of manpower is attributable in the first two communes to the higher multiple-cropping ratio; in the next two units it is accounted for by relatively high land-labor ratios. For the other communes and brigades in the two winter wheat regions, the same measure yields estimates of 170 to 230 days per annum, because of less cultivated land per laborer. For Nan Wei Tzu People's Commune in the spring wheat region, finally, with a growing season of 120 days, the corresponding estimate is an impossible 480 days per labor power per year, due to the questionable labor force estimate.

In summary, then, our calculations suggest that although the agricultural labor force is being utilized more fully than during the 1930's, there may still be absorption problems in the northern plains, even with the new agricultural methods. Of course, we must also mention that in response to direct questions, all officials insisted that there was no labor surplus. Instead, they expressed concern about labor shortage, at least during the transplanting and harvesting seasons.

GRAIN OUTPUT AND CONSUMPTION

The achievement of high and stable yields through adequate irrigation, heavy fertilization, intensive cultivation, more multiple cropping, and other methods required increases not only in farm work but also in other inputs. Such modifications and their yield effects have been discussed elsewhere in this report. What remains to be considered is the return to labor and the changes in the level of living of the rural population that result from them.

Output per Labor Force Member

Buck formed estimates of the grain equivalent of total farm output per man equivalent and per capita of the farm population in 1929-33. Similar measures can be derived for the visited communes by multiplying the yields of all grains per hectare of cultivated grainland with the average number of hectares of all cultivated land (rather than of grainland) per agricultural labor power as well as per capita of the communal population. The results for labor force members are as follows:

Agricultural Region	Grains, kg: Man Equivalent, 1929-33	Grains, kg: Labor Power, 1976	People's Commune or Production Brigade
Spring wheat	787	5,412?	Nan Wei Tzu
Winter wheat-millet	1,112	1,887	Ma Chi Chai
Winter wheat-kaoliang	1,444	2,916	Shuang Chiao
		2,856	Sino-Albanian
		2,781	San Chiao Tsun
		1,562	Huai Ti
		>1,850	Chih Ma
Yangtze rice-wheat	1,357	3,041	Chiang Ning
		3,657	Ma Lu

In all but one of the visited communes, farm labor productivity exceeds previous regional averages substantially, although not necessarily in proportion to the increase in workdays. Agricultural labor is currently least productive in the Huai Ti brigade, which has an unusually low land-labor ratio but also one person per household in outside employment. A similar situation is indicated for Ma Chi Chai People's Commune in the winter wheat-millet region. The contrast for the spring wheat region is once again distorted by the problematic labor force share in Nan Wei Tzu People's Commune.

Output per Capita

The relative differences between the per capita output estimates are greater yet and perhaps more indicative of the actual state of relations, given the problems of labor force definition. Moreover, in this case one also finds meaningful data for Nan Wei Tzu People's Commune, and the situation in Huai Ti Production brigade is uniquely abnormal, like its land-labor ratio:

	Grains per Capita, kg		
Agricultural Region	Farm Population, 1929-33	Commune Population, 1976	People's Commune or Production Brigade
Spring wheat	220	1,300	Nan Wei Tzu
Winter wheat-millet	284	1,254	Ma Chi Chai
Winter wheat-kaoliang	426	729	Shuang Chiao
		743	Sino-Albanian
		770	San Chiao Tsun
		508	Huai Ti
		>845	Chih Ma
Yangtze rice-wheat	483	873	Chiang Ning
		1,153	Ma Lu

Of course, the most surprising result is the negative correlation between the two sets of data in this ordering, which may not be significant. Otherwise it would attest to an extraordinary relevance of the new agricultural methods for climatically less advantageous areas.

Production and Consumption

The greater output per capita, which may yet understate the actual result because nongrain output is not adequately accounted for, did not translate itself into proportionately greater consumption, whether of grains or in general. So far as the disposal of grain is concerned, Buck found for 1929-33 that 25-30 percent of it left the farms, in exchange for cash or as payment in kind to landlords; 55-60 percent was consumed by the farm population, apparently at the average rate of 220-240 kg per capita. Of course, the actual consumption of grain could and did vary greatly from this average with production, by farm size as well as by location.

In comparison, the visited communes supplied their members in 1975 with rations of 220-300 kg of unprocessed grain per capita per annum, irrespective of the size or age and sex structure of the households. These rations, which could be processed and consumed by the households in any way they wished, accounted for 19-44 percent of the (adjusted) per capita output, the unweighted average being 27 percent. The supply of about 0.7 kg of grain per capita per day is considered adequate by all commune officials queried, and it indeed is more than a person usually eats in their opinion. As a result, households accumulate grain inventories, an accumulation viewed as a positive event. Brigades and communes also maintain and build up collective reserves. However, most of the remainder goes for feed and seed if it is not sold to the state.

The receipts from the sale of grains as well as of other products provide the communes with cash income. But the amount of this cash which is distributed to the commune members, as a rule in the form of earning shares in proportion to their work efforts, is also relatively small. In the communes that the Wheat Studies Delegation visited, the total amounts distributed, i.e., income in kind at farm prices plus cash income, constituted 43-62 percent of communal gross receipts in 1975, the average share being 46 percent. The highest share of 62 percent was reported by Huai Ti brigade, which not only had little land and a low product per capita but also experienced opportunities for outside employment that made the development of communal production less decisive.

In the normal case, a larger portion of the total gross receipts is withheld not only to cover costs of production (30 percent on the average), agriculture tax (4 percent on the average), and communal welfare obligations (2 percent on the average), but also to form capital collectively. From 5 to 19 percent of the communal gross receipts (12 percent on the average) were spent for this purpose in the visited communes, *in addition to the 6-30 percent of workdays spent in basic construction.* In contrast, the individual farm household in China traditionally tended to save 2-3 percent of its gross income and engaged to a much more limited extent in capital construction activities. Needless to add, the commune member household is encouraged to continue a few of these practices, i.e., to save individually as well.

IDEAL AND REALITY

The preceding review leads to the conclusion that in the communes that the Wheat Studies Delegation had the opportunity to visit, the Chinese practice farming in a way that utilizes farmers more fully than before and at the same time makes them more productive. As a result they are able to both raise their current level of living and accumulate the means for additional development in the future, including rural industrialization, which has not been discussed as such but which must be the eventual result of further progress.

In the absence of comprehensive macrodata, which the Chinese government chooses not to make public, it is impossible for us to be sure that these methods are similarly effective everywhere. The bits of aggregate evidence put together by various departments of the U.S. Government raise at least some doubts. They imply the following:

1. The area of cultivated land per capita of the rural commune population exceeds 2 mu (= 1/3 acre = 2/15 ha); per agricultural labor force member it probably reaches 6 mu (= 1 acre = 2/5 ha). Farmers on the average thus cultivate and live on more land than those in the visited communes do.

2. Somewhat more than 40 percent of all cultivated land is being irrigated to an unspecified degree, and a larger share of it has been leveled or terraced. These changes are much less complete than the land modification in the visited communes, irrespective of the fact that some land need not be irrigated (or drained).

3. The multiple-cropping ratio, which is unknown, probably exceeds 1.5, but it may well fall short of the magnitude implied by the model practices in the visited communes. Similar differences may be suspected in fertilizer preparation, field management, etc., for which data are likewise lacking.

4. The supply of inorganic fertilizer has risen to the equivalent of about 300 kg of ammonium bicarbonate per hectare of cultivated land on the average, i.e., to less than the amount used per *sown* hectare in the visited communes.

5. The output of all grains per hectare of cultivated land appears to have reached 2,550 kg in 1975, i.e., not much more than one fourth of the average reported by the visited communes. The likely output per agricultural labor force member (probably about 1,000 kg) and per capita of the rural commune population (apparently in excess of 350 kg) account for close to 40 percent of the corresponding averages for the visited communes.

Some of the observations that the Wheat Studies Delegation had the opportunity to make reinforce this contrast. In particular, fleeting vistas from the train en route from Sian via Taiyuan to Shihchiachuang and on to Peking indicated substantial variations in effectiveness. Most of the land had been terraced even well up the slopes of hills, and irrigation facilities were a common sight. Other visible means and methods of cultivation also seemed to be the same as in the communes and brigades that we had inspected. Yet the stands of wheat or barley frequently looked poor. At times, this appeared to be attributable to nutrient deficiencies, at other times to water shortage, and in a few places to alkalinity of the soil.

We could not tell whether these apparent causes of lower yields reflected merely less favorable environmental conditions, which made it more difficult and costly to raise yields, or whether they also resulted from less substantial increases and improvements in efforts, investments, management, and technology. So far as problems of nature are concerned, we knew of course of China's commitment to reduce nutrient deficiencies by raising the production capacity for inorganic fertilizers through the importation of 12 urea plants during the next few years. We also were shown plans to solve water management problems on a much larger scale, e.g., for the entire Hai River system, by means and methods of "basic construction" similar to those seen in the communes. A substantial part of the truly massive earth-moving efforts seems to have been made already. We did not inquire into the problems of treating saline and alkali land.

There is no way of telling whether the work efforts have increased everywhere by as much as in the model communes. Such a change would imply that farm labor is being utilized to a very high degree in most locations. But it also would mean that these incremental efforts are not everywhere highly rewarding and therefore perhaps hard to sustain in the absence of appropriate amounts of fertilizer, water, and other complementary inputs--just as previously during the Great Leap Forward. Similar concerns may be expressed about the quality of farm management, which was evidently excellent in the visited communes but possibly less distinguished elsewhere. With respect to problems of human nature, then, we can merely say that we were shown models which reveal what can be done.

BIBLIOGRAPHY

For the sake of brevity, this report omits references. All the data presented in this section were obtained either during interviews in China or from the following sources:

Buck, J. L. 1930. *Chinese farm economy*. Chicago: University of Chicago Press.

Buck, J. L. 1937. *Land utilization in China*. 3 vols. Nanking: University of Nanking.

Schran, P. 1969. *The development of Chinese agriculture, 1950-1959*. Urbana: University of Illinois Press.

U.S. Congress, Joint Economic Committee. 1975. *China: A reassessment of the economy*. Washington: U.S. Government Printing Office.

U.S. Department of Agriculture, Economic Research Service. 1976. *The agricultural situation in the People's Republic of China and other communist Asian countries*. Foreign agricultural report 124.

9

SUMMARY AND CONCLUSIONS

Wheat, as China's second most important grain crop, occupies an estimated area of approximately 26 million ha, of which 85 percent is winter wheat. According to the Chinese, more than 80 percent of its wheat is irrigated. The Honan, Shantung, Hopei, Anhwei, Shensi, and Szechwan provinces are China's largest producers of wheat.

The Chinese identify eight wheat-producing regions. The *Yellow River-Hwai River Plains region* produces weak winter and spring types. Spring wheats and strong winter types are grown in the *northern region*. Springs and strong winters also are grown in the *northwestern region*, but spring wheats only can be grown in the *northeastern region*, where the winter months are dry and cold. Fall-seeded spring wheats and weak winters are found in the *middle and lower reaches of the Yangtze River*, but only spring seeding of spring wheats occurs in the *upper reaches of the Yangtze River and Szechwan Basin*, where weak winter wheats also are grown. Fall- or winter-seeded spring wheats are grown in the *southern China region* except for some spring seeding at the higher elevations. Only winter wheats are found on the *Tibet Plateau*.

China is rich in genetic diversity of wheat, some of which was observed by the Wheat Studies Delegation at Wukung. China initiated a program in 1955 to systematically collect, catalog, and preserve wheat cultivars, but we were unable to fully ascertain the current status of the collection. We were told that germ plasm banks have been established, but responsibility for their maintenance is assigned to the provinces. A reliable assessment of China's wheat germ plasm preservation effort was not possible on the basis of information provided and material observed.

Broad genetic diversity was observed in the crossing blocks of breeding programs at some institutes and academies. Winter and spring wheats from several foreign countries were being utilized as parents of new hybrid combinations. Semidwarf spring varieties from Mexico were observed in nurseries in northeast China, but the disposition and whereabouts of substantial amounts of experimental materials sent to China by CIMMYT in the last 3 years were not determined.

Chinese agricultural leaders were reluctant to engage in discussions of systematic exchanges of wheat germ plasm with the United States. China does not cooperate or participate in any of several existing international wheat evaluation networks. Team members repeatedly emphasized the mutuality of benefits of Chinese participation in germ plasm exchanges and international evaluation networks. A substantial amount of wheat germ plasm from the United States was brought to China by the Wheat Studies Delegation. The Chinese reciprocated by presenting seed of 70 Chinese varieties to the U.S. delegation on its departure from China.

The major objectives of all wheat-breeding programs visited are earliness, to enhance the use of wheat in multiple-cropping systems, and development of water responsive semidwarf cultivars adapted to irrigation. Breeding wheats for rain-fed production was rarely mentioned by the Chinese. Resistance to diseases, particularly scab, also is a major objective of the breeding programs visited. The most comprehensive wheat-breeding programs were at the province level, involving various agricultural research institutes and academies of agricultural science. Conventional hybridization in single, three-way, and double crosses followed by pedigree selection was most commonly practiced, although some mutation breeding and pollen culture techniques were observed. There was no evaluation of selections for processing quality and protein content in programs visited.

The overall impression gained from a 4-week exposure to wheat breeding in China was a highly positive one. The programs visited at the province level were impressive, and the breeders were capable people. The breeding programs conducted at the commune, brigade, and production team levels were serving an effective extension role in bringing scientific methods to the farmers. How effective as breeding activities the programs are at the commune and subcommune levels is questionable. There appears to be much duplication of effort at these levels, but as extension functions, they are effective.

Wheat cultivars currently in commercial production in China obviously are high-yielding types that have good resistance to the most serious diseases and are responsive to irrigation and good management practices. Some lodging was apparent under very high fertility levels. Owing to the nature and diversity of the environment of China and the presence of many diseases, there will be a continuing need for new sources of resistance. As increasingly higher yield levels are attained and greater intensity of production realized, it can be anticipated that increasingly complex production problems also will be encountered; problems that seemingly will require new basic information for solution.

The Chinese have developed an apparently highly effective agricultural extension system for dissemination of new information and new wheat cultivars, which is accomplished by the continuing close cooperation and interchange between farmers and scientists and a team approach to wheat improvement. Training in plant breeding seems to be mainly through practical experience gained from working closely with older scientists. Emphasis is totally on applied technology to solve current production problems. Genetic research in wheat was lacking at the institutes and colleges visited. The Wheat Studies Delegation was impressed with wheat breeding in China and breeding accomplishments to date. It failed to find the backup research in genetics, physiology, statistics, and other basic sciences required for continued breeding advances. This could seriously impede future agricultural progress in China.

Plant pathology and entomology are combined into one discipline in China, identified as plant protection. A very large number of people (probably more than 50,000) are engaged in plant protection activities, which are largely an application of information known since 1950. Protection is achieved through development of resistant varieties via breeding and through clean seed and plant quarantine administered at the county and provincial levels. Plant protectionists at the commune and subcommune levels were young, bright, enthusiastic people, only minimally acquainted with fundamental concepts of pathology and entomology. Lack of training and knowledge of basic concepts also were apparent at the provincial institutes and academies.

Head blight and scab caused by *Gibberella zeae* is the most important wheat disease in China and apparently the only one not yet under control. Marked differences in scab infection were observed, suggesting that the Chinese wheats may possess some of the best known resistance to scab. Powdery mildew (*Erysiphe graminis*) is China's second most important wheat disease. Aphid species as vectors of barley yellow dwarf probably are the most important foliar wheat insects in China, but such soil insects as armyworms, stem maggots, and wireworms cause sporadic damage and may be the most important wheat insects overall. There is a question as to whether the massive trial-and-error approach to wheat protection based on existing knowledge can effectively cope with current and future wheat problems of China.

The intercropping and multiple cropping of wheat with other crops appear to have been well researched in China, but apparently by experience more than by replicated experiment. Chinese knowledge of multiple cropping may be unsurpassed in the world. Despite the overwhelming importance of water policy and irrigation to Chinese agriculture and wheat production, the Wheat Studies Delegation did not encounter experiments on irrigation, plant-water relations, water use efficiency, or effects of moisture stress on plant metabolism. Weeds were not a problem in wheat fields examined, but control was mainly by hand removal rather than use of herbicides. Organized basic crop physiology research relating to photosynthesis, photorespiration, or related processes was not observed. Use of replication and sound experimental design in wheat field experimentation appears to have been largely abandoned, the suggestion being that technicians have received minimal or no training in design of experiments.

The diversity of climatic and ecological features of China is expressed by an unusually wide range of soil conditions. Communes visited were mainly located on benchlands of valleys or coastal plains where soils were formed from alluvial or loess materials. Wheat production on soils of lower quality on sharply sloping land of northern Shensi and Shansi provinces and on sandy deposits of the Wei River as well as on saline and alkali spots in valleys was observed from the train between Sian and Peking. Extensive land shaping has occurred or is in progress throughout visited wheat areas of China to accommodate gravity flow irrigation.

China may be the world's champion of organic manuring of soil. All organic wastes including human wastes are carefully husbanded. Green leguminous manures play a significant role in supplying nitrogen.

Emphasis on short-term practical soil research of a demonstrational kind, rather than long-range research, may have serious implications for China's future soils and fertility technology. Plowing is considered a primary and essential tillage practice for loosening the seedling zone and incorporation of copious quantities of organic manures.

Use of chemical fertilizers on wheat is very low in relation to use of organic fertilizer, this despite substantial experimental evidence demonstrating that yields can be raised significantly by the use of chemical fertilizer. Manufacturing capability for chemical fertilizers is currently under construction, which, when completed, will make China the world's second largest producer. This additional fertilizer, together with improved water control and utilization, offers excellent prospects for continued increases in wheat production in China.

The Chinese have developed a simple but effective system for storage of grain in which sound, well-planned facilities, low grain moisture content, and constant monitoring of grain condition are key components. The delegation was highly impressed with the level of sanitation in and around storage facilities and the excellent control of insects and rodents. Milling of wheat into flour for baked goods and noodles is done at the subcommune level and in much larger urban flour mills. Extraction rates of flour are much higher than in the United States. There is little, if any, attention given to maximizing the nutritional value of milled products.

The Chinese identify Liberation in 1949 and the Cultural Revolution, which began in 1966, as major turning points in the recent history of China. Chinese society appears to have been highly politicized, with the masses of people mobilized to a degree that was unprecedented prior to 1966. Observed uniformity and pervasiveness of politicization surpassed expectations of delegation members. The delegation was administered a heavy dosage of political rhetoric at each institute, college, factory, and commune visited, which followed essentially the same format at each place. The Chinese apparently acquire the rhetoric through constant reiteration and involvement in frequent small-group study sessions.

The basic structure of agricultural research and extension work in China involves seven levels. The highest four levels--central government, provinces, prefectures, and hsien--are operated by the state whereas the bottom three levels--communes, production brigades, and production teams--are operated by the people collectively.

A close relationship has been forged between agricultural scientists and the rural masses in what the Chinese call "conducting scientific research in an open-door way"; an approach that demands that scientific research be conducted through constant and close cooperation of scientists and peasants. Barriers between intellectuals and manual laborers are further broken down through so-called three-in-one combination as the backbone of production and agricultural successes.

The leading personnel of all administrative units visited by the Wheat Studies Delegation are organized into revolutionary committees, with party members seeming to dominate. The Chinese now follow a policy of sending educated city youth to live and work in rural communes as a method of reducing the size of bulging cities and raising the educational level of the rural areas. These policies have clarly improved the lives of farmers and have altered agricultural research.

The Wheat Studies Delegation was favorably impressed by the results of most of these policies as evidenced by apparently well-fed, healthy farmers; intelligent, informed, and committed political leadership; and seemingly vigorous, enthusiastic agricultural scientists. Heavy stress upon *current* production problems together with *radical* implementation of the open-door approach to agricultural research could have significant future negative effects, however.

The Chinese believe that agricultural productivity and farm income can be substantially increased by greater effort even in the absence of modern production methods. The model communes visited by the Wheat Studies Delegation demonstrated the practicality and validity of this approach. Greater effort is accomplished by basic construction to bring all or most of the land under irrigation, preparation of enormous quantities of organic fertilizer, increasingly intensive cultivation of each crop, and multiple cropping of land to the limits of climate.

The result has been increased total farm output and increased output per unit of land, per farm laborer, and per capita of population. Commune members in 1975 were provided rations of 220-300 kg of unprocessed grain per person. Sale of the remainder of grain production and other products provides the communes with cash income. Farming-team incentives are maintained by the communes through improved peasant diets, gradually increased household incomes, better social services, and the rebuilding of family residences.

Aware that the changes brought about by the Cultural Revolution have been many and that the time to absorb these changes has been short, the delegation is cautious in making any summary comments on the Chinese academic system as it relates to agricultural education. It would be wise to wait for several more classes of students to pass through the system before judging the quality of education that they have received; however, it is possible to discern the outlines that appear to be guiding the restructuring of the education system, and the delegation does feel it would be useful to their American colleagues to comment on these outlines.

The Chinese educational structure appears to be producing competent technicians trained in specific, narrow fields. The priorities in the training of agricultural specialists are directed at increasing immediate crop yields and solving problems related to crop production. Classes have a high component of practical experience in them, the teaching staff being not only traditionally trained professors but also experienced peasants, members of the People's Liberation Army, and other students. Much time is spent in the fields in what is termed research but is essentially extension work.

The delegation does have some concerns with some aspects of the educational system as it now appears to be functioning. One of the chief concerns is that students seem to lack training in dealing with problems of experimental design. The students lack training in statistical analysis and do not use replicated experiments to check or confirm experimental results. Soil chemistry, soil physics, and soil microbiology do not seem to be subjects taught at the current time. Plant pathology is devoted to the immediate present and not to the problems of the future. Such problems as the effects of intercropping and double cropping, high irrigation rates, high fertilizer rates, and the use of spring wheats in north China--practices planned for the near future--are not being adequately researched or taught. With the exception of some work being done at the Kiangsu Provincial Agricultural Science Research Institute, there seems to be little effort devoted to crop physiology as it relates to photosynthesis or photorespiration.

Another concern of the delegation was recruitment of teaching staff for agricultural colleges. It seems that the practice of keeping on some of the graduates of each university to become teachers could well lead to problems of stagnation, problems that have, in the past, been evident in some U.S. universities. It was the observation of the delegation that any teaching faculty needs to be stimulated by new ideas generated by new people brought into the teaching program, ideas that result from different educational programs and different research fields or topics. At present, this does not seem to be the case.

The delegation was uniformly impressed with the quality and enthusiasm of the students whom they met in all parts of China. Their genuine desire to work for the betterment of all Chinese people was evident, and it is clear from the increased production figures in food grains that their goals of increasing yields in wheat are being met. However, the delegation felt that these short-term goals might well be obviated if in the long run the students are not provided with more of the basic crop science tools that they will be able to use when confronted with problems that cannot now even be foreseen.

It should be borne in mind that the educational system in China is in a state of flux and movement; many of the scientists who met with the delegation were aware of the current problems of the system and were working to overcome them. The delegation felt that future CSCPRC delegations should continue to observe and comment on the evolving nature of Chinese agricultural education so that a more complete and accurate picture of both the process of this evolution and the final result can be achieved.

APPENDIX A

COMMITTEE ON SCHOLARLY COMMUNICATION WITH THE PEOPLE'S REPUBLIC OF CHINA: WHEAT STUDIES DELEGATION

ROBERT H. BUSCH
Associate Professor
Department of Agronomy
North Dakota State University
Fargo, North Dakota 58102

Dr. Busch's main research interests are in quantitative inheritance of economic characteristics of spring wheat and more effective methods of effecting genetic control and selection.

R. JAMES COOK
Research Plant Pathologist and Laboratory Leader
Regional Cereal Disease Research Laboratory
U.S. Department of Agriculture
Associate Professor of Plant Pathology
Washington State University
Pullman, Washington 99165

Dr. Cook's main research interests are in ecology of root-infecting fungi in soil and in the host plant rhizosphere, especially possible mechanisms of use to biological control.

LLOYD E. EASTMAN
Professor of History and Asian Studies
Department of History
University of Illinois
Urbana, Illinois 61801

Dr. Eastman's main research interests are in Chinese history from 1937 to 1949 with special emphasis on the period of anti-Japanese war and the 4 years prior to the establishment of the People's Republic of China.

VIRGIL A. JOHNSON, *Chairman*
Supervisory Research Agronomist
U.S. Department of Agriculture, Agricultural Research Service
Professor of Agronomy
University of Nebraska
Lincoln, Nebraska 68583

Dr. Johnson's main research interests are in wheat breeding and genetics, genetics and physiology of wheat, protein quantity, and nutritional quality.

WARREN E. KRONSTAD
Professor of Agronomy
Oregon State University
Corvallis, Oregon 97331

Dr. Kronstad's main research interests are in cereal improvement, environment-genotype interaction, use of biometrical models to partition genetic variation, disease resistance, influence of chelating agents on genetic recombination, and aluminum tolerance in plants.

DALE N. MOSS
Professor of Crop Physiology
Department of Agronomy and Plant Genetics
University of Minnesota
St. Paul, Minnesota 55108

Dr. Moss's main research interests are in effects on photosynthesis, respiration, and transpiration of higher plants of light intensity, temperature, carbon dioxide concentration, nutrition, removal of storage organs, air turbulence, planting patterns, and age of leaves.

RAMON H. MYERS
Scholar-Curator
East Asian Collection
Hoover Institution
Stanford University
Stanford, California 94305

Dr. Myers' main research interests are in agricultural development and rural problems of East Asia with emphasis on China, the Chinese farm economy before the Cultural Revolution, and the organization of farming under socialism since the Cultural Revolution.

MR. KOY NEELEY
Foreign Agricultural Service
U.S. Department of Agriculture
United States Liaison Office
Peking, China

Mr. Neeley is the agricultural officer attached to the U.S. Liaison Office in Peking.

ROBERT A. OLSON
Professor of Agronomy
University of Nebraska
Lincoln, Nebraska 68583

Professor Olson's main research interests are in plant nutrition studies with radioisotope tracers, improving validity of soil testing, and abatement of agricultural pollution.

YESHAJAHU POMERANZ
Director
Grain Marketing Research Center
U.S. Department of Agriculture
1515 College Avenue
Manhattan, Kansas 66502

Dr. Pomeranz's main research interests are in chemical composition and functional properties of cereal grains, nutritional value of cereals, high-protein foods, food rheology, and microbiology of stored grain products.

ALAN P. ROELFS
Research Plant Pathologist
U.S. Department of Agriculture
Cereal Rust Laboratory
University of Minnesota
St. Paul, Minnesota 55108

Dr. Roelfs' main research interests are in cereal rust epidemiology and physiological race distribution, resistance of *Berberis* and *Mahonia* species to *Puccinia graminis*, and losses caused by the cereal rust diseases.

PETER SCHRAN, *Secretary*
Professor of Economics and
Asian Studies
Department of Economics
University of Illinois
Urbana, Illinois 61801

Dr. Schran's responsibilities include those of a staff member of the CSCPRC. His main interests are in the rural aspects of economic development and in Chinese policies of rural development.

APPENDIX B

ACTUAL ITINERARY

May 17 Monday	Arrival in Tokyo for 2 days of preparatory meetings.
May 19 Wednesday	Travel by plane from Tokyo to Peking. Reception at the airport by Mr. Hsü Yün-t'ien, member of the board of directors of the Agriculture Association (AA), and other representatives of this organization.
May 20 Thursday	Meeting with Mr. Wang Chên-ko, office chief of the AA, and other representatives to discuss and agree on the itinerary. Visit to the Imperial Palace Museum. Visit to the Peking Institute of Genetics of the Chinese Academy of Sciences. Banquet given by Mr. Li Yung-k'ai, director general of the AA.
May 21 Friday	Visit to the Peking Institute of Botany of the Chinese Academy of Sciences. Visit to the Peking Institute of Atomic Energy Use in Plant Breeding of the Chinese Academy of Agricultural and Forestry Sciences. Lloyd Eastman and Ramon Myers visit bookstores in Peking.
May 22 Saturday	Visit to Shuang Chiao People's Commune in the suburbs of Peking. Briefing by Chairman Chuang, inspection of wheat fields, flour mill, and research laboratory. Lecture by Mr. Hsü Yün-t'ien on the development of wheat production in China. Meeting presided over by Mr. Li Yung-k'ai. Attendance of performance by Canton acrobats.
May 23 Sunday	Visit to Great Wall and to Ming tombs. Picnic at Chang Ling.

May 24 Monday	Flight from Peking to Nanking. Reception by Fêng Hsün, responsible member of the Kiangsu branch of the AA. Discussion of local itinerary. Visit to Kiangsu Provincial Agricultural Sciences Research Institute. Inspection of plant protection laboratory and wheat experiment fields, followed by long exchange of views on various aspects of wheat breeding. Banquet given by Mr. Fêng Hsün.
May 25 Tuesday	Sight-seeing in Nanking. Visit to the Yangtze Bridge and to Hsüan Wu Lake. Visit to the Institute of Soil Sciences of the Chinese Academy of Sciences. Reception by Professors Hsiung I and Li Ch'ing-k'uei (part of the delegation returned for further discussions to the Agricultural Sciences Research Institute). Attendance of *Song and Dance* performance.
May 26 Wednesday	Visit to Chiang Ning People's Commune. Briefing by Chairman Yang. Inspection of wheat fields, experimental fields, agricultural research station, grain store, flour mill, noodle shop, farm machinery works, and hospital.
May 27 Thursday	Messrs. Cook, Johnson, Kronstad, and Moss meet with approximately 40 scientists from six institutes in Kiangsu for lectures on objectives in American wheat breeding; photosynthesis with respect to wheat; breeding of high-yield, semidwarf, rust-resistant varieties; and control of scab and related diseases. (The rest of the team visits the Nanking Museum of Revolutionary History meanwhile.) Visit to the Sun Yat-sen Memorial and to the Meteorological Institute (Observatory). Flight from Nanking to Shanghai. Reception by Mr. Ch'en Yu-hua, vice-director of the Shanghai branch of the AA. Discussion of local itinerary.
May 28 Friday	Visit to Shanghai Institute of Plant Physiology (Messrs. Eastman, Myers, and Schran visit the Chao-Yang Neighborhood.) Visit to Shanghai Academy of Agricultural Sciences. Inspection of experimental fields, laboratories, and library. Banquet given by Tung Chia-p'ing, director of the Shanghai branch of the AA.
May 29 Saturday	Visit to Ma Lu People's Commune in Chia-ting Hsien, Shanghai. Briefing by Vice-Chairman Li. Inspection

	of wheat fields, irrigation works, laboratories, workshops, hospital, and housing facilities.
May 30 Sunday	Visit to Shanghai Industrial Exhibition with subsequent sight-seeing. Visit by Messrs. Busch, Cook, Johnson, Kronstad, Olson, and Roelfs to Shanghai Agricultural Exhibition. Visit by Messrs. Eastman, Moss, Myers, Neeley, Pomeranz, and Schran to the Shanghai Flour Mill.
May 31 Monday	Flight from Shanghai via Nanking and Chengchow to Sian. Reception by Mr. Ch'ang Chün-i, vice-director of the Shensi branch of the AA. Discussion of itinerary with Mr. Hsü Ju-chou, secretary-general of the Shensi branch of the AA. Visit to Shensi Province Historical Museum. Banquet given by Ch'ang Chün-i.
June 1 Tuesday	Visit to Northwest Agricultural College in Wukung, Shensi Province. Inspection of experimental fields, laboratories, and library. Prolonged discussions in disciplinary groups.
June 2 Wednesday	Visit to Hua Ching Hot Springs. Visit to Ta Hsing Lou municipal grain storage unit. Visit to Sian Clock Tower and to Big Goose Pagoda.
June 3 Thursday	Visit to Shensi Province Academy of Agricultural and Forestry Sciences in Wukung. Inspection of germ plasm fields, lectures by Messrs. Cook, Johnson, Kronstad, Olson, and Roelfs.
June 4 Friday	Visit to Ma Chi Chai People's Commune outside Sian. Briefing by Chairman Hsin. Inspection of wheat fields, experimental station, subsidiary enterprises, and flour mill. Visit to Panpo Archaeological Museum. Departure for Shihchiachuang via Taiyuan, by train.
June 5 Saturday	Arrival in Shihchiachuang. Reception by Hua Chien, responsible person of the Hopei branch of the AA. Discussion of local itinerary. Visit to Hai River control exhibit. Attendance of *Song and Dance* performance by young red guards.
June 6 Sunday	Visit to San Chiao Ts'un Production Brigade of Yung An People's Commune and to Wu Hsing Production Brigade of Wu Hsing People's Commune, both of Cheng Ting Hsien. Discussion of the four-level research system with

responsible persons of the Cheng Ting Hsien research unit.
Visit to Huai Ti Production Brigade of Huai Ti People's Commune in the outskirts of Shihchiachuang.
Inspection of fields and flour mill.
Banquet given by Hua Chien.

June 7 Monday	Visit to Chih Ma People's Commune in Luan-cheng Hsien. Study of interplanting of commercial crops. Visit to Hopei Cereal Crop Research Institute. Inspection of experimental fields and laboratories. Discussions.
June 8 Tuesday	Visit to Shihchiachuang Martyr Garden with tomb of Dr. Norman Bethune. Return to Peking by train.
June 9 Wednesday	Briefing by Pao Wen-k'uei and Sun Yuan-shu of the Peking Academy of Agricultural Sciences on triticale research in China. Meeting presided over by Li Yung-k'ai. Visit to Sino-Albanian Friendship People's Commune outside Peking. Inspection of triticale and wheat fields, duck farm, and discussion of communal organization and operation. Visit with Mr. Thomas, economic councillor of the U.S. Liaison Office in Peking.
June 10 Thursday	Visit to Summer Palace. Flight from Peking to Changchun plus drive from Changchun to Kungchuling. Reception by Mr. Chao Ta, chairman of the board of directors of the Kirin branch of the AA. Discussion of local itinerary. Banquet given by Mr. Chao Ta.
June 11 Friday	Visit to Kirin Academy of Agricultural Sciences. Inspection of germ plasm and of experiments in breeding and intercropping. Prolonged discussions with the academy's staff.
June 12 Saturday	Visit to Nan Wei Tzu People's Commune in Huai Te Hsien. Inspection of wheat fields and of the experiment station of Big Elm Tree Production Brigade. Visit to Kirin Academy of Agricultural Sciences. Lectures by Messrs. Busch and Kronstad with subsequent discussion.
June 13 Sunday	Trip to Erh Shih Chia Tzu (20-Families) People's Commune in Huai Te Hsien, Kirin Province. Visit of an exhibit on rusticated educated youth in the commune and of a collective household of such youths. Return from Kungchuling to Changchun.

June 14 Monday	Visit to the school of the Changchun acrobatic troupe and to South Lake Park. Flight from Changchun via Shenyang to Peking.
June 15 Tuesday	Meeting with representatives of the AA. Presentation of gifts of germ plasm and publications by Chin Shan-pao, vice-chairman of the AA. Discussion of delegation members' questions by Mr. Hsü Yün-t'ien. Lectures by Messrs. Johnson and Kronstad to approximately 40 Chinese scientists. Meeting attended by Messrs. Busch, Cook, Moss, Olson, and Roelfs. Visit by Messrs. Eastman, Myers, and Schran to the Peking Revolutionary Museum. Farewell banquet given by the delegation for Li Yung-k'ai, other members of the AA, and other hosts in the Peking area.
June 16 Wednesday	Sight-seeing and shopping in Peking. Flight from Peking to Tokyo.
June 17 Thursday	Final meeting of the delegation in Tokyo.

APPENDIX C

PERSONS MET IN CHINA

The names of persons met by the delegation during our visit are presented in the order of our itinerary, with one modification: All the individuals met in Peking are listed in the beginning, even though we saw some of them for the first time during our second and third stays in that city. All names are listed by organizational affiliation in the order in which they were given to us, which usually reflects their ranking. Names without characters could not be checked for spelling.

Properly at the head of the list appear the names of our four Chinese travel companions, who did so much to make our tour enjoyable as well as instructive.

李登春 Mr. Li Teng-ch'un
Chinese Academy of Agricultural and Forestry Sciences: Wheat Breeder

黄永宁 Mr. Huang Yung-ning
Chinese Association of Agriculture: Secretary

楊 鴻 Ms. Yang Hung
Chinese Association of Agriculture: Interpreter

黄毫信 Mr. Huang Hao-hsin
Chinese Association of Agriculture: Interpreter

Peking

Chinese Association of Agriculture

金善宝 Mr. Chin Shan-pao
Vice-Chairman

李永凯 Mr. Li Yung-k'ai
Director General

王枕戈 Mr. Wang Chên-ko
Member of the Board of Directors
Office Chief

許运天 Mr. Hsü Yün-t'ien
Member of the Board of Directors
Wheat Breeder

黄永宁 Mr. Huang Yung-ning
Secretary

趙錫麟 Mr. Chao Hsi-lin
Staff Member

張云千 Mr. Chang Yün-ch'ien
Staff Member

潘紹宗 Mr. P'an Shao-tsung
Interpreter

楊 鴻 Ms. Yang Hung
Interpreter

黄堯信 Mr. Huang Hao-hsin
Interpreter

王幼琼 Mr. Wang Yu-ch'iung
Interpreter

徐錦华 Ms. Hsü Ching-hua
Interpreter

Chinese Academy of Sciences

Institute of Genetics

胡 含 Mr. Hu Han
Vice-Chairman of the Revolutionary Committee

刘桐华 Mr. Liu T'ung-hua
Director of the Technical Division

庄家駿 Mr. Chuang Chia-chun
Geneticist (pollen)

胡啟德 Mr. Hu Ch'i-te
Geneticist (remote hybridization of wheat)

李向輝 Mr. Li Hsiang-hui
Geneticist (somatic hybridization)

段声付 Mr. Tuan Sheng-fu
Representative of the Poor and Lower Peasants

袁开文 Mr. Yuan K'ai-wen
Interpreter

Institute of Botany

李森 Mr. Li Sên
Responsible Person of the Revolutionary Committee

崔澂 Mr. Ts'ui Ch'êng
Plant Physiologist

王世之 Mr. Wang Shih-chih
Wheat Breeder

王恩玉 Mr. Wang Ên-yü
Secretary

梁松筠 Mr. Liang Sung-yün
Taxonomist

郭仲琛 Mr. Kuo Chung-ch'ên
Cytologist

冯晋庸 Mr. Feng Chin-yung

朱至清 Mr. Chu Chih-ch'ing

荆玉祥 Mr. Ching Yü-hsiang

王春茂 Mr. Wang Ch'un-mao

刘存德 Mr. Liu Ts'un-tê

Chinese Academy of Agricultural and Forestry Sciences

Atomic Energy Utilization Research Institute
(Institute of Atomic Energy)

趙文璞 Mr. Chao Wen-p'u
Responsible Person

刘書城 Mr. Liu Shu-ch'êng
Research Office

王林清 Mr. Wang Lin-ch'ing
Breeding Office

崔聰淑 Ts'ui Ts'ung-shu
Breeding Office

温賢芳 Wen Hsien-fang
Radiation Laboratory

江素心 Chiang Su-hsin
Radiation Laboratory

尤宗杓 Mr. Yu Tsung-shao
Physicist and Chemist

黄日模 Mr. Huang Jih-mo
Physicist and Chemist

Peking Academy of Agricultural Sciences

姚宗文 Mr. Yao Tsung-wen
Chairman of the Administrative Office

鮑文奎 Mr. Pao Wen-k'uei
Senior Triticale Breeder

孫元樞 Mr. Sun Yuan-shu
Triticale Breeder

Sino-Albanian Friendship People's Commune

马宝庭 Mr. Ma Pao-t'ing
Vice-Chairman

張永和 Mr. Chang Yüng-ho
Experiment Station
Staff Member

楊基录 Mr. Yang Chi-lu
Responsible Person for Production

周書秾 Mr. Chou Shu-nung
Technician

刘炳亮 Mr. Liu Ping-liang
Officer

KIANGSU PROVINCE

Nanking

Association of Agriculture--Kiangsu Branch

馮迅 Mr. Fêng Hsün
Responsible Person

卢良恕 Mr. Lu Liang-shu
Deputy Secretary-General
Wheat Breeder

吴纪华 Mr. Wu Chi-hua
Member of the Board of Directors

马榮棠 Mr. Ma Jung-t'ang
Member of the Board of Directors

Association of Friendship with Foreign Countries--Kiangsu Branch

汪光森 Mr. Wang Kuang-shen
Representative

Kiangsu Province Agricultural Sciences Research Institute

楊致平 Mr. Yang Chih-p'ing
Vice-Chairman of the Revolutionary Committee

阮德成 Mr. Yung De-cheng
Responsible Person in the Research Office

周朝飞 Mr. Chou Zhao-fei
Wheat Breeder

梅藉芳 Mr. Mei Gi-feng
Wheat Breeder

刘大钧 Mr. Liu Da-chung
Wheat Breeder

崔继林 Mr. Zhei Chi-ling
Physiologist of Rice

沈梓培 Mr. Sing Hsin-pie
Soil Man Scientist

陆培元 Mr. Lu
Pathologist

过崇儉 Mr. Gu Chung-chiang
Pathologist

王裕中 Mr. Wang Yu-chung
Plant Protectionist (scab)

范衡标 Mr. Fang Hing-piao
Staff Member
Member of the Revolutionary Committee Office

Institute of Soil Research

熊毅
Mr. Hsiung I
Vice-Chairman of the Revolutionary Committee
Physicist and Chemist

李庆逵
Mr. Li Ch'ing-k'uei
Vice-Chairman of the Revolutionary Committee
Agricultural Chemist

龚子同
Mr. Kung Tzu-t'ung
Soil Geographer

刘铮
Mr. Liu Chêng
Agricultural Chemical Lab--Microelement Group
Specialist for Microelements

刘文政
Mr. Liu Wen-chêng
Specialist for Saline and Alkaline Soils

俞仁培
Ms. Yü Jen-p'ei
Specialist for Saline and Alkaline Soils

马一杰
Mr. Ma I-chieh
Soil Physicist and Chemist

Chiang Ning People's Commune

楊立生
Mr. Yang Li-sheng
Chairman of the Revolutionary Committee

胡加云
Mr. Hu Chia-yun
Vice-Chairman of the Revolutionary Committee

吕界超
Mr. Lü Chieh-ch'ao
Vice-Chairman of the Revolutionary Committee

王庆和
Mr. Wang Ch'ing-ho
Party Secretary of Chiang Ning Town of Chiang Ning County

吕桂祥
Mr. Wu Kuei-hsiang
Head of the Agricultural Technique Station of the Commune

Shanghai

Association of Agriculture--Shanghai Branch

陳玉华 Mr. Ch'en Yu-hua
Deputy Secretary-General

楊禹臣 Mr. Yang Yü-ch'ên
Staff Member

張光祖 Mr. Chang Kuang-tzu
Staff Member

朱福生 Mr. Chu Fu-shêng
Staff Member

Shanghai Foreign Affairs Office

陳定平 Mr. Ch'ên Ting-p'ing
Staff Member

Shanghai Institute of Plant Physiology

沈允纲 Mr. Shen Yun-kang
Professor, Photosynthesis

湯章城 Mr. T'ang Chang-ch'eng
Responsible Person for Phytotron

陳敬祥 Mr. Ch'en Ching-hsiang
Member of the Revolutionary Committee
Vice-Chairman of the Research Team

Kwang Yin-jiang
Responsible Person for Plant Hormone and Weed Control

Wang Ming-yang
Deputy Director of the Administrative Office

Wei Jia-miang
Researcher of Photosynthesis

Shanghai Academy of Agricultural Sciences

儲昕 Mr. Ch'u Hsin
Responsible Person for the Scientific Research Group

Chi Lu-tiang
Responsible Person of the Revolutionary Committee

Li Pu-ta
Responsible Member of the Research Office

Chen Tei-sung
Responsible Person for Crop Breeding and Cultivation of Crops

Yu Chin-hung
Responsible Person for Soil, Fertilizer, and Plant Protection

Li Heng
Breeding and Cultivation of Crops

Ma Jin-fu
Breeding and Cultivation of Crops

Sung Wei-gen
Breeding and Cultivation of Crops

Wang Fang-tao
Soil and Fertilizer and Plant Protection

Shie Hsu-chen
Soil and Fertilizer and Plant Protection

Ma Lu People's Commune

李承訓 Mr. Li Ch'êng-hsün
Vice-Chairman of the Revolutionary Committee

徐浩新 Mr. Hsü Hao-hsin
Responsible Member of the Plant Protection Station

陳星根 Mr. Ch'ên Hsing-kên
Responsible Person for Scientific Research Work

彭永添 Mr. P'êng Yung-t'ien
Staff Member

張峯 Mr. Chang Fêng
Staff Member

Shanghai Flour Mill

繆必兴 Mr. Miao Pi-hsing
Chairman of the Revolutionary Committee

刘文彬 Mr. Liu Wên-pin
Vice-Chairman of the Revolutionary Committee

范存明 Mr. Fan Ts'un-ming
Director of the General Office

趙鴻鳴 Mr. Chao Hung-ming
Member of the Revolutionary Committee

SHENSI PROVINCE

Association of Agriculture--Shensi Branch

常君毅 Mr. Ch'ang Chün-i
Vice-Director

許汝周 Mr. Hsü Ju-chou
Secretary-General

刘隨倉 Mr. Liu Sui-ts'ang
Staff Member

党明孝 Mr. Tang Ming-hsiao
Staff Member

Shensi Province Travel Service

王国堂 Mr. Wang Kuo-t'ang
Interpreter

Northwest College of Agriculture (Wukung College of Agriculture)

董 巩 Mr. Tung Kung
Vice-Chairman of the Revolutionary Committee

李志才 Mr. Li Chih-ts'ai
Vice-Director of the Revolutionary Committee's General Office

趙洪章 Mr. Chao Hung-chang
Professor
Wheat Breeder

張海峯 Mr. Chang Hai-fêng
Lecturer of Wheat Breeding

楊天章 Mr. Yang T'ien-chang
Lecturer of Wheat Breeding

荊家海 Mr. Ching Chia-hai
Lecturer of Plant Physiology

马鸿运 Mr. Ma Hung-yüng
Lecturer of Agricultural Economics

李振岐 Mr. Li Chêng-ch'i
Associate Professor of Plant Pathology

李生秀 Mr. Li Sheng-hsiu
Lecturer of Soil Science

夏魯平 Mr. Hsia Lu-p'ing
Staff Member

吴中录 Mr. Wu Chung-lu
Interpreter

賈文林 Mr. Chia Wen-lin
Interpreter

万沾中 Mr. Wan Chan-chung
Interpreter

王广森 Mr. Wang Kuang-shêng
Interpreter

路进生 Mr. Lu Chin-shêng
Interpreter

Shensi Province Academy of Agricultural and Forestry Sciences

楊生海 Mr. Yang Shêng-hai
Vice-Chairman of the Revolutionary Committee

蔚向秀 Mr. Wei Hsiang-hsiu
Responsible Person of the Revolutionary Committee

張頤 Mr. Chang I
Responsible Person of the Cereal Crops Research Division

宁琨 Mr. Ning K'un
Wheat Breeder

王玉成 Mr. Wang Yü-ch'êng
Wheat Breeder

許志魯 Mr. Hsü Chih-lu
Wheat Breeder

Mr. Chu Hsiang-san
Responsible Person for Plant Protection Research Division

路端宜 Mr. Lu Tuan-i
Wheat Rust Researcher

刘汉文 Mr. Liu Han-wên
Plant Pathologist

李德润 Mr. Li Tê-jun
Responsible Person of the Soil and Fertilizer Research Division

王云梯 Mr. Wang Yün-t'i
Researcher of the Organic Manure Research Division

張学上 Mr. Chang Hsüeh-shang
Responsible Person of Green Manure Research Division

Sian Ta Hsing Lou Grain Storage Unit

張宗仁 Mr. Chang Tsung-jên
Director

王寿山 Mr. Wang Shou-shan
Vice-Director

潘仁云 Mr. P'an Jên-yün
Vice-Director

吴崇斌 Mr. Wu Ch'ung-pin
Staff Member

侯永生 Mr. Hou Yung-shêng
Staff Member

王宣 Mr. Wang Hsüan
Staff Member

朱则一 Mr. Chü Tsê-i
Staff Member

Ma Chi Chai People's Commune

辛向尧 Mr. Hsin Hsiang-yao
Chairman of the Revolutionary Committee

娄志兴 Mr. Lou Chih-hsing
Vice-Chairman of the Revolutionary Committee

田忠科 T'ien Chung-k'o
Vice-Chairman of the Revolutionary Committee

王庆庄 Mr. Wang Ch'ing-chuang
Staff Member

HOPEI PROVINCE

Association of Agriculture--Hopei Branch

华 践 Mr. Hua Chien
Responsible Person

刘学怀 Mr. Liu Hsueh-huai
Staff Member

侯永才 Mr. Hou Yung-ts'ai
Staff Member

Hopei Office of Relations with Foreign Countries

范永年 Mr. Fan Yung-nien
Responsible Person

Hopei Cereal Crops Research Institute

孙景福 Mr. Sun Ching-fu
Vice-Director

郭秀峯 Mr. Kuo Hsiu-fêng
Vice-Director

刘福昌 Mr. Liu Fu-ch'ang
Scientist of Wheat Division

黎迁华 Mr. Li Chian-hua
Scientist of Wheat Division

龚邦鐸 Mr. Kung Pang-to
Scientist of Wheat Division

王士奎 Mr. Wang Shih-k'uei
Scientist of Culture Technique Division

Hopei Plant Protection and Soil Fertility Institute

卢福瑞 Mr. Lu Fu-jui
Vice-Director

胡明峻 Mr. Hu Ming-chün
Scientist of Insect and Disease Section

付秋舫 Mr. Fu Ch'iu-fang
Scientist

孫全先 Mr. Sun Ch'üan-hsien
Scientist

Yung An People's Commune, San Chiao Ts'un Production Brigade

賈焕辰 Mr. Chia Huan-ch'ên
Vice-Chairman of the Revolutionary Committee of Cheng Ting County

張五岳 Mr. Chang Wu-yüeh
Vice-Chairman of the Revolutionary Committee of Cheng Ting County

王連全 Mr. Wang Lien-ch'üan
Staff Member of Cheng Ting County's Agriculture Research Unit

杜 鈌 Mr. Tu Ch'üeh
Staff Member of Cheng Ting County's Agriculture Research Unit

范宝林 Mr. Fan Pao-lin
Vice-Chairman of Yung An Commune's Revolutionary Committee

王志新 Mr. Wang Chih-hsin
Vice-Chairman of San Chiao Ts'un Brigade's Revolutionary Committee

Wu Hsing People's Commune

刘兰瑞 Mr. Liu Lan-jui
Vice-Chairman of the Commune's Revolutionary Committee

于順祥 Mr. Yü Shun-hsiang
Chairman of Wu Hsing Production Brigade's Revolutionary Committee

黃太仲 Mr. Huang T'ai-chung
Agricultural Technician of Wu Hsing Production Brigade

Huai Ti People's Commune, Huai Ti Production Brigade

刘天保 Mr. Liu T'ien-pao
Chairman of the Commune's Revolutionary Committee

孔令為 Mr. K'ung Ling-wei
Chairman of the Brigade's Revolutionary Committee

孔秋來 Mr. K'ung Ch'iu-lai
Vice-Chairman of the Brigade

李风玲 Ms. Li Feng-ling
Director of the Woman's Association

蔣群球 Mr. Chiang Ch'ün-ch'iu
Agricultural Technician

Chih Ma People's Commune, Tung Yang Shih Production Brigade

潭士英 Mr. T'an Shih-ying
Vice-Chairman of the Revolutionary Committee of the Hsien

周壬午 Mr. Chou Jên-wu
Vice-Director of the Administrative Office of the Hsien

田喜井 Mr. T'ien Hsi-ching
Vice-Chairman of Chih Ma Commune

韓德發 Mr. Han Tê-fa
Vice-Chairman of Tung Yang Shih Brigade

王 浩 Mr. Wang Hao
Staff Member

何志偶 Mr. Ho Chih-ou
Staff Member of the Agricultural and Forestry Bureau of the Hsien

王志輝 Mr. Wang Chih-hui
Agricultural Technician of the Hsien

王炳辰 Mr. Wang Ping-ch'ên
Agricultural Technician of the Brigade

KIRIN PROVINCE

Association of Agriculture--Kirin Province Branch

趙述 Mr. Chao Ta
Chief of the Board of Directors

马相榮 Mr. Ma Hsiang-jang
Member of the Board of Directors

Kirin Academy of Agricultural Sciences (KAAS)

張一夫 Mr. Chang I-fu
Vice-Chairman of the Revolutionary Committee

朱天林 Mr. Chu T'ien-lin
Vice-Director of the Administrative Office

王进先 Mr. Wang Chin-hsien
Wheat Breeder

白金凱 Mr. Pai Chin-k'ai
Plant Pathologist

武克忠 Mr. Wu Ko-chung
Specialist of Crop Cultivation

KAAS Institute of Crop Breeding

李公德 Mr. Li Kung-tê
Vice-Director
Sorghum Breeder

時俊峯 Mr. Shih Chün-fêng
Vice-Chairman

張希昌 Mr. Chang Hsi-ch'ang
Vice-Director of the Administrative Office

Nan Wei Tzu People's Commune

宋 玉 — Mr. Sung Yü
Chairman of the Commune

牛国太 — Mr. Niu Kuo-t'ai
Chairman of the Brigade

APPENDIX D

BRIEF ACCOUNTS OF THE VISITED UNITS

During its tour of the previously listed organizations of research and production, the U.S. Wheat Studies Delegation received at the beginning of each visit a briefing on the work of the visited unit. The brief accounts of the 10 research institutes or academies and the one college may be summarized as follows:

INSTITUTE OF GENETICS OF THE ACADEMY OF SCIENCES, PEKING

The institute, which was visited on May 20, 1976, did not exist before Liberation, when work on genetics was carried on only in universities. Its predecessor was the Genetic Selection Laboratory, which was organized in 1951 with a staff of 20 persons. This laboratory was transformed into the present institute in 1958, when its staff was increased to 100 persons, who worked primarily in plant genetics. Since the Cultural Revolution, the institute has developed greatly, its staff having grown to more than 200 researchers.

Presently, the institute operates four laboratories plus one experiment center. The molecular genetics laboratory works on leukemia and other medicine-related problems. The animal genetics laboratory deals with developmental genetics, transplantation, and sex control. The plant genetics laboratory conducts research in pollen culture, somatic hybridization, and inheritance characteristics, mostly of rice and potatoes. The fourth laboratory studies male sterility in higher plants (in particular, sorghum) and remote hybridization (of cotton). The experiment station, located not in Peking but on Hainan Island, experiments with mutagenesis through chemical and physical induction.

Since the Cultural Revolution, much of the research work is being carried on outside the institute "in an open-door way," especially by three methods: (1) Workers and peasants are invited to join the research work in the institute and to share with the staff their practical experience; at the time of the visit there were 18 peasants in such positions, for periods of up to 2 years. (2) Institute staff go out to factories, communes, and hospitals to help in the solution of practical problems; one third of the staff go each year, for 2 to 3 months during the growing period. (3) The institute cooperates with other units (e.g., hospitals) in efforts to make research problem oriented and thus "serve the people."

The most important achievement mentioned concerned the shortening of the period required for the development of new stable varieties of rice and wheat, which used to be 8 years. By the use of new methods (pollen culture) it has become possible to accelerate the process. Work on a new variety of wheat was started in 1970 and completed in 1971.

INSTITUTE OF BOTANY OF THE ACADEMY OF SCIENCES, PEKING

The institute, which was visited on May 21, 1976, grew out of previously existing units, predominantly the Peiping Botanical Institute, with a combined staff of 30 persons at the time of Liberation, who engaged primarily in the survey and taxonomy of plants. The institute was established in its present form in 1950. Its staff has grown to 300 persons, less than 50 percent of whom are women. Its research agenda has broadened considerably, especially since the Cultural Revolution.

The work of the institute is organized in seven laboratories. The plant taxonomy laboratory has undertaken an inventory of the flora of China, to be published in an 80-volume edition, five volumes of which have been completed; it also compiled a four-volume edition of commercially useful plants, which was presented to the delegation. The plant ecology laboratory deals with problems of environmental protection of vegetation. The paleobotany laboratory conducts fossil research in connection with oil exploration. The plant chemistry laboratory studies commercial plants (herbs, oil-bearing plants, and gelatines). The work of the plant physiology laboratory is subdivided into hormone research, storage of vegetables and fruits, herbicides, photosynthesis, and wheat studies; the latter involve the study of close planting and the development of an index of optimal spacing. The cytology and morphology laboratory conducts work on pollen culture and cytohybridization. The nitrogen fixation laboratory focuses on leaf surface fixation.

The close-planting study started in 1958, during the period of the Great Leap Forward and commune formation. Since the Cultural Revolution, research is being carried on in an "open-door way" through the four-level research network in the people's communes, and it has become more practical and production oriented. Noted examples were the publication of illustrated pamphlets on various plants, especially medicinal herbs.

INSTITUTE OF ATOMIC ENERGY USE IN PLANT BREEDING OF THE ACADEMY OF AGRICULTURAL AND FORESTRY SCIENCES, PEKING

Preparations for the establishment of the institute began in 1957, and operations started in 1960. The institute is subdivided into a political office, a scientific research office, and an administrative office. The scientific research office has three laboratories: the laboratory for plant breeding with radiation, which breeds new varieties of wheat, sorghum, maize, and cotton through radiation and hybridization; the isotope-tracing laboratory, which was established in 1961; and the physics and chemistry laboratory, which undertakes research and experimental construction of radiation instruments and agricultural instruments. In addition, the institute operates an experimental farm, an experimental workshop, and experimental bases in people's communes.

Since the Cultural Revolution, one third of the staff go to the countryside each year to help the peasants. The practical research problem that they try to solve is how to increase the coefficient of multiple cropping by making it possible to transplant larger plants.

KIANGSU PROVINCIAL AGRICULTURAL SCIENCES RESEARCH INSTITUTE, NANKING, KIANGSU PROVINCE

The institute, which was visited on May 24, 1976, is the oldest institute in China. It operates seven divisions: food grain studies, commercial crops, plant protection, soil and fertilization, animal husbandry, vegetables and fruits, and physics and agronomy (radiation). It also has 1,000 mu of experimental fields and a staff of 600 persons.

The principal research objective is to change the cropping system so as to increase the yield of all crops. In wheat breeding, this implies efforts at achieving earlier (5-7 days) maturity, greater disease resistance and fertilizer receptivity, and higher protein content. Methods used to achieve these results are a mass movement in seed selection (natural variation); the crossing of different varieties; breeding by radiation (cobalt 60), which produced Ningmai no. 3; and the study of early maturity genetics, with the local ecology as a starting point.

Since the Cultural Revolution, this work is related to the mass movement in research and tied in with the four-level research network. Research units have been set up in all hsien, 80 percent of the people's communes, 60 percent of the brigades, and an unspecified share of the teams. In Kiangsu Province, 1.2 million people have joined in the research work. The institute helps them to set up such units and projects.

To this end as well as to further its own research, the institute exchanges researchers with units in the field. It has organized four comprehensive teams with distinctive cooperative affiliations. Two thirds of the institute's staff are out in rural areas, half of them permanently and half of them temporarily. Only one third of the staff work in the institute at any particular time. The latter number is augmented by peasants with a certain level of education and practical experience who are invited to join for some time. In addition, the institute cooperates with other units in research projects concerning practical problems of Kiangsu Province.

Regular additions to the staff are requested from and assigned by the Kiangsu Department of Higher Education, and most of them are persons from Kiangsu. The experienced peasants are selected and sent by the people's communes.

INSTITUTE OF SOIL SCIENCE OF THE ACADEMY OF SCIENCES, NANKING, KIANGSU PROVINCE

The institute, which was visited on May 25, 1976, was set up in 1951.

It now has a staff of 450 persons. Its predecessor before Liberation was the Division of Soils of the Geological Survey of China, which was concerned with the mapping of soils.

The institute is subdivided into eight research divisions: soil geography; soil physical chemistry; soil agrochemistry; soil biochemistry; microorganisms; soil physics; improvement of saline soil; and soil pollution. It also maintains a library, greenhouse, experimental facilities, and workshops. The library has 80,000 volumes and is up to date in foreign journals but 10 or more years behind in monographs.

The principal research activities are a soil (utilization) survey, studies of how to improve low-yield land (red soil, saline soil, and alkaline soil), studies of how to improve the soil quality of high and stable yield fields, studies of the relation between fertilizer and soil, studies of soil micro-organisms, and studies of soil pollution.

Since the Cultural Revolution, more attention is being paid to production, and research is carried out in an open-door way. Some new activities have been undertaken: A soil survey is being made for northwest China, northeast China, and Tibet; experimental centers have been established in north and south Kiangsu as well as in south China for the study of red soils and in north China for the study of alkaline and saline soils, in cooperation with the peasants; do-it-yourself soil analysis kits have been prepared and supplied to each of the 1,800 people's communes in Kiangsu for use by the commune research centers. Stocks of chemicals are maintained by the hsien research center staffs.

INSTITUTE OF PLANT PHYSIOLOGY OF THE ACADEMY OF SCIENCES, SHANGHAI

The institute, which was visited on May 28, 1976, was set up in 1953 with a staff of 23 persons. By now, the staff has grown to 431 persons, including 275 research workers, 44 percent of whom are women. The institute is subdivided into six departments: cell physiology; biological nitrogen fixation; photosynthesis; phytohormones; microbiology; and phytotron. The institute's phytotron, a very impressive and versatile unit, was completed in 1960. It was designed and built in China, has controlled temperature and humidity and contains 25 controlled environment rooms with 360 m^2 of floor space. It is the largest phytotron in China.

The method is to combine with the peasants and masses in open-door way research. Cadres and scientists go to the villages and stay for a long time in order to learn from the peasants. Experienced farmers are invited to the academy to participate in research work in three-in-one combinations (workers-cadres-scientists). In addition, cooperative work is carried out with more than 100 units, including people's communes, state farms, hospitals, factories, schools, and universities, in order to promote production. Finally, the academy is involved in discussions of production with others and in the extension of its research results to other units.

NORTHWEST COLLEGE OF AGRICULTURE, WUKUNG, SHENSI PROVINCE

The college, which was visited on June 1, 1976, was founded in 1934. From Liberation (May 20, 1949) until the Cultural Revolution (May 16, 1966), the school was run on the old foundation. The "Liu Shao-ch'i line" dominated in teaching, with the result that the 5,225 graduates acquired a revisionist orientation and did not like to work in agriculture. During the Cultural Revolution the working class came to the college and took over. Bourgeois intellectuals do not rule anymore; the school is open to workers and peasants and soldiers, who now constitute 95 percent of the student body of 2,171 persons.

The college is run by a revolutionary committee consisting of a three-in-one combination of teachers, students, and cadres. Teaching is carried on in an open-door way. More than 1,000 students and more than 300 teachers go out to study and struggle in the field. More than a thousand experimental centers have been set up in the five northwestern provinces. Six hundred and thirty students have worked at three experimental centers in the Kuanchung region and in south Shensi.

The college conducts regular courses of 1- to 3-years duration plus shorter training courses according to the demands of production. More than a thousand persons have been trained in such short-term courses since 1971. Since 1975, regular students are required to return to the communes from which they came. One thousand two hundred and sixty-nine workers, peasants, and soldiers have been enrolled under this program since its initiation, and more than 600 of them will graduate in 1977 and return to their places of origin.

The organizational structure, teaching program, and research activities were not described in detail for the delegation as a whole. However, indications of the work that is being done were given in group discussions under five headings: wheat breeding, soils, pathology, physiology, and agricultural economy.

SHENSI ACADEMY OF AGRICULTURAL AND FORESTRY SCIENCES, WUKUNG, SHENSI PROVINCE

The academy, which was visited on June 3, 1976, was started in 1952 and began research in 1954. It experienced several name changes until 1973, when it acquired its present designation. The academy is organized into 10 sections, the first seven of which are located in Wukung, about 80 km west of Sian. The sections are food grains and crops, plant protection, animal husbandry, forestry, special plant research, vegetables, soil and fertility, cotton, fruit trees, and silkworms. The academy employs a staff of 1,300 persons, including 700 cadres. It cultivates 3,000 mu of experimental fields and is the principal germ plasm depository for the northern wheat provinces.

Since the Cultural Revolution, the academy conducts its work in an open-door way and tries to learn from Tachai in its experiments. It operates more than 100 experimental centers in 70 hsien and carries out 480 research projects, which may be summarily described as stressing breeding for high yields; intercropping; plant protection; optimal fertilization; "learn from Tachai" in soil management; collecting, preserving, and utilizing germ plasm of the major crops; and collecting fruit tree strains as well as silkworm lines.

The academy has been successful in 130 of these projects. It undertakes extension work concerning its research results as well as its scientific techniques. More than 100 books and articles have been domestically issued. Agricultural science is being popularized through broadcasting and by other means. More than 13,000 persons have been trained in the more than 100 centers, which set examples for the training by the prefectures and hsien (of more than 400,000 persons in 1975). Seventy percent of the academy's research activities are carried out through the four-level network in hsien, people's communes, etc. In addition, the academy engages in socialist cooperation with other institutions inside and outside the province on more than 30 projects.

HOPEI CROP RESEARCH INSTITUTE, SHIHCHIACHUANG, HOPEI PROVINCE

The institute, which was visited on June 7, 1976, was set up in the early 1950's in the western part of Hopei. It moved to its present location in 1972 but is still in the process of constructing its research facilities. Its germ plasm bank continues to be located at the original site, the laboratories are temporary structures, and the library remains crated.

The institute has a staff of 230 persons, 40 percent of whom are involved in scientific work, and it cultivates 950 mu of experimental fields. It is subdivided into seven divisions: wheat research, cropping systems, corn, millet, cereal protection, pollen culture, and soybeans and oil-bearing plants. In addition, there are laboratories plus an administrative office.

The principal research activities of the institute concern the creation of good and improved varieties that are high yielding and disease resistant and the study of the effects of good cultivation practice on yield, in the light of the "eight-point charter" and the requirements of Hopei Province.

Since the Cultural Revolution, the institute has carried on research in an open-door way, learning from Tachai, etc. It operates more than 40 research stations in people's communes. One third of the staff go every year to work in the communes, and the internal work of the institute is integrated with the work in the communes. In addition, experienced peasants are invited to participate in the work of the institute.

KIRIN ACADEMY OF AGRICULTURAL SCIENCES, KUNGCHULING, KIRIN PROVINCE

The academy, which was visited on June 11, 1976, was set up in 1921 by the Japanese as an agricultural experiment station. Its name was changed in 1949 to agricultural science research institute and again in 1955 to its current designation. Presently, the academy has 805 employees, including 259 research workers, 436 ordinary workers, and 110 administrative staff. It cultivates 6,500 mu of land, including 1,500 mu of experimental fields and 5,000 mu for feed crops and seed multiplying.

The academy's work is organized in five divisions: crop and breeding research; soil and fertilizer, cropping systems, and cultivation; plant protection; animal husbandry; and fruit trees. The main tasks are to solve key problems of provincial agricultural production, to popularize extension, and to train personnel. With respect to wheat, the work of the first three divisions involves breeding new varieties that are adaptable to irrigation, fertilizer tolerant, high yielding, and early maturing; developing interplanting, catch cropping, and multiple cropping; and coping with wheat diseases and insects, especially rusts and armyworms.

Since the Cultural Revolution, this work is integrated with the four-level research network. More than 60 percent of the staff go to units in this system. In 1976, they maintained more than 30 research bases and more than 500 liaison units. Breeding work is carried on together with 22 of these bases in people's communes, through cooperative agreements. Intercropping is being developed in cooperation with five hsien units. Forecasting of armyworm plagues is done through more than 300 observer stations in people's communes.

TEN RURAL PEOPLE'S COMMUNES

The information received in introductory briefings on and subsequent discussions of the work of the visited communes was sufficiently quantitative and homogeneous to merit tabulation. Because of the specific interests of the delegation, however, these brief accounts did not always follow the standard format, so that numerous gaps appear in the following 16 tables. The visited units are identified by number in the order of the delegation's itinerary as follows:

No.	Production Brigade	People's Commune	Hsien (County)	Nearby Location
1		Shuang Chiao		Peking
2		Chiang Ning	Chiang Ning	Nanking, Kiangsu
3		Ma Lu	Chia-ting	Shanghai
4		Ma Chi Chai		Sian, Shensi
5	San Chiao Ts'un	Yung An	Cheng Ting	Shihchiachuang, Hopei
6	Wu Hsing	Wu Hsing	Cheng Ting	Shihchiachuang
7	Huai Ti	Huai Ti		Shihchiachuang
8		Chih Ma	Luan-cheng	Shihchiachuang
9		Sino-Albanian Friendship		Peking
10		Nan Wei Tzu	Huai Te	Kungchuling, Kirin

The eleventh unit, Erh Shih Chia Tzu People's Commune in Kirin Province, is omitted because the delegation's visit yielded an insufficient amount of comparable data.

TABLE D-1 Structure and Population

No.	Production Brigades	Production Teams	Member Households	Member Population	Birth Rate per mill	Death Rate per mill	Rusticated Youths
1	6	62	8,400	40,000	12[a]	6[a]	2,500
2	14[b]		7,810	30,922	15.2	6.1	300[c]
3	14	144	7,020	28,600	14[d]	4	600
4	23	81	4,540	22,200	12	5-6	130[e]
5		9	640	2,840	13	7.6	
6		16	1,190	5,550	5.5?	3.3?	
7		8	918	3,917	12	5	
8	9	70	>4,000	19,100	9	4	>300
9	28	71	6,800	32,000	10[f]	2.5	900
10	12	106	4,500	23,000	8	2	<600

[a] In 1966, 30 births per mill, 10 deaths per mill.
[b] Plus one fishing team.
[c] Five hundred came, 200 have left.
[d] Before Cultural Revolution, 20 births per mill.
[e] Two hundred came, 70 have left.
[f] Before Cultural Revolution, 28 births per mill.

TABLE D-2 Labor Force and Employment

No.	Labor Force Members[a]	Communal Employment: Agriculture[b]	Communal Employment: Industry	Communal Employment: Other	Employed in Outside Work	Average Annual Workdays per Communal Laborer
1	17,600	10,000	4,000			300
2	11,000	90%	1,050	800[c]?		300-320
3	18,000[d]	85%	15%	>200[e]	2,000?	300-310
4	11,000[f]		400	500[g]	4,000	
5	>1,000					
6						
7	1,400	90%	10%		>1,000	310
8	9,000		410			
9	>12,000	<85%[h]	700			
10	6,130?	90%	10%		<150	310

[a] The reference is to lao-tung li (labor powers).
[b] Including subsidiary work activities.
[c] Handicrafts: 400; commerce: 200; transport: 200.
[d] Twelve thousand full labor powers, 55 percent of them women.
[e] Handicrafts.
[f] Sixty percent full labor powers.
[g] Commerce.
[h] Less than 70 percent in crop growing, 1,400 in animal husbandry.

TABLE D-3 Education

	Number of Schools					May 7 University		
No.	Primary	Middle	Total[a]	Students	Teachers	Type	Students	Students Sent to Regular College
1	19	6		14,000		Yes[b]	Yes[b]	
2	41	5		6,200	262			
3			24	5,000	700	Spare Time		6
4	16	1		5,300	190			5
5			1	600	15			
6								
7								
8	9	6			400	Spare Time	210	
9	21	5		>8,500	600	Work-Study	120	300[c]
10	12	13		5,000	300			

[a] Only available data for these two communes were aggregate data on total number of schools.
[b] Data given delegation only confirmed existence of both May 7 universities and May 7 university students.
[c] Total since 1969.

TABLE D-4 Medical Care

No.	Hospitals	Hospital Beds	Hospital Staff	Medical Stations	Barefoot Doctors	Cost of Medical Care,[a] yuan
1						1
2	1	50	48	14	46	2
3	1	30	49	14	59	2
4	1		80	23	85	
5				1	7[b]	
6						
7						
8	1	10	>11[c]	9	53	
9	1		>20[c]	28	142-213[d]	
10	1		35	12	36[e]	

[a]Per capita per annum.
[b]Plus two regular doctors.
[c]Number of doctors.
[d]Two to three per team.
[e]Plus 100-200 persons in public hygiene.

TABLE D-5 Territory and Land Use

No.	Gross Area, km^2	Culti- vated Area, ha	Irrigated Area, share	Land, ha Grains	Rice	Wheat	Other	Coefficient of Multiple Cropping
1	90	3,600	Most		2,000	1,400		1.5-2.0
2	50	2,787	Most					>2.2
3	30	2,261	Most	50%			35%[a]	2.6
4	37.5?	1,767	98%			1,067		1.6-1.7
5		193				115		1.7
6		463						
7		170						1.85
8		2,141		1,340[b]		1,314		
9	50	3,780	90%	2,100		1,667		1.8
10	80	4,531	Little[c]		1,000	100	1,500[d]	1.0?

[a]Cotton. Plus 15 percent in oil seeds, fruits, and vegetables.
[b]Cultivated land.
[c]Transition to irrigation to be completed in 3 years.
[d]Soybeans.

TABLE D-6 Animal Stock

No.	Pigs	Chickens	Ducks	Cattle	Horses
1	Yes[a]		Yes[a]	Yes[a]	
2	20,000				
3	45,517[b]				
4	12,000	25,000		210	
5					
6					
7	3,000[c]			None	
8	19,000			None	
9	32,000		56,000	2,100	1,400
10	11,500			400	2,500

[a]Only data received by delegation at this commune was that it did have pigs, ducks, and cattle.
[b]In 1949, 4,170; 1957, 6,780; 1965, 34,101.
[c]Including 2,300 pigs in brigade pens.

TABLE D-7 Farm Machinery

No.	Large Tractors	Hand Tractors	Total Tractors	Trucks	Pumps	Other Equipment
1	50	150				6[a]
2	11	174		5		1,143[b]
3		200			37	Many
4			80	4	660	Cutters
5						
6						
7	1	13		2		
8					260	1,077
9	50	110			185[c]	>300[d]
10	53	19			30	

[a]Harvester combines. Other equipment not mentioned.
[b]Including 2 bulldozers, 166 diesel and 388 electric motors, and 587 threshers as well as pumps.
[c]One hundred more planned, 60 of them completed.
[d]Including 19 earthmovers, 9 combines, 30 large and 125 small threshers, 125 winnowers, 3 well sinkers, etc.

TABLE D-8 Work Requirements

No.	Man-Days of Cultivation per ha[a]		Man-Days of Fertilizer Work per ha	Share of Basic Construction in All Days
	Wheat	Cotton		
1	240-300[b]			
2	300-375[c]			15%
3	300[d]	375-450	375[i]	10-25%
4	300-450[e]			5%
5	>300		120	Small[j]
6				
7	>450[f]	>750	15-30	10%
8	210-225	300-315		20%
9	<150[g]		90	30%
10	450[c,h]			30%

[a] Exclusive of fertilization.
[b] On high-yield plots.
[c] Same as rice.
[d] Similar to rice and corn.
[e] Same as corn.
[f] Transplanting of wheat.
[g] High degree of mechanization.
[h] Including fertilization.
[i] Moving silt at 2.5 man-days per cubic meter.
[j] Forty to fifty percent of all days during 1965-66.

TABLE D-9 Fertilization and Chemical Treatment

No.	Input of Organic Fertilizer		Input of Inorganic Fertilizer		Price, yuan/kg	Yield Response Ratio	Cost of Insecticide or Soil Treatment, yuan/kg
	Quantity, tons/ha	Man-Days, days/ha	Quantity, kg/ha	Type of Ammonium			
1	60-75		500-750	Sulphate	0.32	1:3-1:5	3.00-4.50
2	135		300	Bicarbonate	0.18	1:6?	
3	150	375		Sulphate	0.26		7.50
4	150		150	Sulphate?	0.26		1.50
5	75-112	120		Sulphate	0.34	1:4	
6							
7	150	15-30	375	Bicarbonate	0.16	1:2-1:3	
8	75-90		375[a]	Sulphate?	0.33[b]		
9	75	90	750	Bicarbonate	0.18		
10	80		100[a]	Bicarbonate	0.22		

[a]Plus an equal amount of phosphate.
[b]Average, 0.32-0.34 yuan/kg.

TABLE D-10 Irrigation

No.	Amount of Water,[a] m^3/ha	Cost of Water,[a] yuan/ha	Number of Appli- cations	Source of Water	Depth of Wells, m	Source of Power
1		20[b]	5-6	Wells	90	Electric
2						
3		9.00				
4		10.50	3	Wells		Electric
5[c]		3.00	12-13	Wells		Electric
6						
7	675	5.25	2-3	Wells		Electric
8	510-525	3.00	Several	Wells		Electric
		9.00-10.50				Diesel
9	450	13.50	4-5	Wells	15	Electric
10		3.00	2	Liao River	4	

[a] Per application.
[b] Cost per annum.
[c] One well per 6 ha.

TABLE D-11 Output of All Grains Per Unit of Cultivated Land, kg/ha

No.	Pre-Liberation	1957	1965	1966	1970	1971	1974	1975
1	800							8,100
2	2,250		6,000			7,500		9,705
3	3,360	4,612	11,407					14,415
4	1,500			5,618				8,085
5				7,500[a]				11,325
6								9,600
7	2,715							11,715
8	1,500		3,240	3,975	6,270			>7,500
9	800		4,065				5,081	6,300
10								6,488

[a]This output level was maintained from 1966 through 1974.

TABLE D-12 Yield of Wheat Per Unit of Sown Land, kg/ha

No.	1965	1969	1970	1971	1972	1974	1975	1976
1							4,000	
2	1,793		2,000				3,000	
3							5,000[a]	
4							5,250	
5	3,000						6,795	
6				3,038		4,350	4,650	
7[b]							6,345	
8		3,000			4,350	4,500	4,500	
9	2,500						3,500	
10							2,250	3,000

[a] Another reference to 350 kg/mu equals 5,250 kg/ha.
[b] In 1949, 1,650 kg/ha.

TABLE D-13 Yields Per Unit of Land Sown to Other Crops, kg/ha

No.	Rice 1975	Barley 1975	Cotton 1949	Cotton 1966	Cotton 1975	Vegetable 1949	Vegetable 1966	Vegetable 1975
1	6,000							
2		3,015[a]						
3		4,000						
4			225	500	750		62,500	75,000
5				>750	1,020			
6					773			
7						>30,000		75,000
8								
9								
10	6,800							

[a] Ch'ing Ch'u Production Brigade, 1974.

TABLE D-14 Production and Distribution of Grains and Other Products

No.	Production: Wheat, tons	Production: Grains, tons	Sale of Grains, tons	Per Capita Ration of Grains, kg	Grain Reserve, tons	Sale of Other Products, tons
1	6,000	18,000[a]	4,000	350[b]		
2	>5,000	26,750[c]	12,900[d]	300[e]	8,360	
3		14,000		268		
4		6,500[f] ?	1,625[g]	240	110[h]	*A*
5		2,410	280	250	515	*B*
6						
7			None	225		
8			2,960	258	650	*C*
9	5,500		3,200[i]	220	3,000	*D*
10			45-48%	260[j]		

A Thirty-one thousand tons of vegetables; 4,300 pigs; 101.5 tons of cotton.
B Six hundred and thirty tons of cotton.
C Three thousand pigs.
D Fifteen thousand tons of vegetables; 500 tons of fruits; 40 tons of fish; 7,000 tons of milk; 56,000 ducks.

[a] Wheat plus rice.
[b] Residual of wheat plus rice.
[c] In 1965, 16,870 tons; 1963, 13,370 tons.
[d] Including more than 4,000 tons of wheat. Sales were twice the quota.
[e] Including about 30 kg of wheat.
[f] Value of food grain production divided by farm price of grains.
[g] In 1966, 335 tons.
[h] A 1975 addition.
[i] Including 2,000 tons of wheat. The 1975 quota was 1,100 tons.
[j] Plus 15 kg of soybeans.

TABLE D-15 State Procurement Prices

No.	Wheat, yuan/kg	Other Products	Price of Other Products, yuan/kg
1	0.20?[a]		
2	0.262-0.272[b]		
3	0.272	Rice	0.22
		Watermelon	0.10
4	0.272		
5	0.31		
6			
7	0.26	Vegetables	0.06
8	0.26	Corn	0.18
		Kaoliang	0.14
		Cotton	2.00
9	0.28	Duck	2.20
10		Rice	0.30
		Soybeans	0.26-0.28

[a]Urban retail price of wheat flour, 0.36 yuan/kg; of rice, 0.28-0.42 yuan/kg.

[b]Wholesale price, 0.226 yuan/kg.

TABLE D-16 Distribution of Income

	Fraction of Gross Value of Communal Production								
No.	Production Cost, %	Agriculture Tax, %	Cost Plus Tax %	Public Accumulation, %	Public Welfare, %	Accumulation Plus Welfare %	Distributed to Members, %	Distribution per Capita per Annum, yuan	Average Monthly Wages of Nonagricultural Labor, yuan
1								130	35-45
2		4.2[a]		9.0				140	40[b]
3	38.5	4.0		12.9	1.5		43.1	200[c]	Work points
4	31.4	3.3[h]		12.4	2.0		51.0	110[d]	45-70
5	40.0	1.7				12.9	44.3	111	Work points[e]
6									
7	20.5	3.8		18.7	2.0		55.0[f]	168?	Work points[e]
8			33	5	2		60	93	
9	43.0	4.5		9.4[g]	2.0		41.1	117	
10	25	5		20	?		45-50	128	Work points

[a] Fixed in 1964, paid in cash.
[b] Highest wage, 70 yuan.
[c] In 1957, 52 yuan.
[d] Average per household: 1975, 560 yuan; 1966, 434 yuan.
[e] Persons working on the outside earn 50-60 yuan per month.
[f] Another statement gave 62 percent versus 38 percent for all other uses.
[g] Largely residential housing construction.
[h] Four hundred eighty-five tons of grain.

APPENDIX E

QUESTIONS TRANSMITTED TO MR. HSÜ YÜN-T'IEN ON JUNE 9, 1976

The U.S. Wheat Studies Delegation was pleased to have Mr. Hsü Yün-t'ien's thorough briefing on the conditions of wheat production in China, which prepared us well for our visits to numerous institutes and people's communes in Nanking, Shanghai, Sian, Shihchiachuang, and Changchun. Unfortunately, it was impossible at the time of the briefing to answer the many questions that arose in our minds in response to this excellent presentation. A large part of them were resolved in discussions with researchers, cadres, and peasants during the course of our tour. However, a few questions remain, and several new ones have arisen as a result of our studies. We present these questions below with the request that they be made the basis for our discussion with Mr. Hsü and other responsible members of the Agriculture Association at the end of our tour of various centers of research and locations of wheat production. We group our questions under five headings:

A. BREEDING AND GENETICS

1. In the Nanking-Shanghai area, where spring wheat varieties are used, we have heard much about the speed of development of new varieties. It has been indicated that this is possible in only 3 years. We would like to know how breeding is organized to permit a breeder to propagate as many as three generations of wheat per year? How much breeding material can he manage this way? How are experimental lines evaluated for specific and general adaptation in so short a time?

2. Do all breeders manage their breeding populations in a similar manner and is generally the same system followed to make new varieties available to the communes? Is there an organized system of regional, provincial, and central evaluation of new experimental lines prior to their increase and distribution to communes for production? Are provincial and central meetings held for dissemination of new research data? How often are they held?

3. How are new wheat germ plasm and new research information disseminated among communes, counties, and provinces? Does there exist a printed listing of wheat varieties grown on communes and state farms in China in which their relative importance and their regions of production are shown?

4. What is the normal educational background of wheat breeders? How do they receive the necessary specialized training for such activities as genetic studies, mutation induction, cell fusion, and anther culture?

B. PLANT PROTECTION

1. *Tilletia controversa* Kühn (TCK) requires very special environmental conditions to infect wheat. These conditions are that the wheat must be winter wheat overwintered beneath snow for many weeks in unfrozen soil. A frozen soil surface or an uncovered soil surface prevents infections. The ideal conditions for such infections occur in many places in the Pacific Northwest but nowhere else in the United States. Where in China do such environmental conditions occur, permitting the infection of wheat by TCK?

2. How is research information concerning new diseases and insects and potential disease epidemic forecasting disseminated among communes, counties, and provinces?

C. IRRIGATION AND SOIL MANAGEMENT

1. What are the relative amounts of irrigation from stream diversion and from groundwater sources in China, and has a significant decline in the water table level occurred on the Peking Plain where pump irrigation is so extensive?

2. How much of inorganic nitrogen fertilizer, phosphorus fertilizer, and potassium fertilizer is now produced in China, and what preparations are being made for absorbing the large quantity of fixed N that will come from the 10 large nitrogen plants now being built?

3. What chemical carriers of N and P are considered to be superior for Chinese soil conditions, and what is the reason for this superiority?

4. What surveillance is being accomplished of nitrate-nitrogen accumulation in groundwater? Has methemoglobinemia in babies from excess nitrate in water or leaf foods been observed to any significant degree?

5. What tillage/residue management practices are employed for the specific objectives of moisture conservation and organic maintenance in soils where dry farming is practiced? With increasing use of heavy machinery in soil tillage are problems with soil compaction being experienced, and if so, how are they being alleviated?

D. PROCESSING QUALITY

1. Does China have central, provincial, or local standards for wheat (and other grains) as well as for wheat products such as flour, bread, and biscuits?

2. Does China have standard methods of testing these products?

3. Who specifies these substantive and procedural standards at the central, provincial, or local level?

4. Who carries out the prescribed tests at these levels, including tests of exported and imported commodities?

5. May the U.S. Wheat Studies Delegation obtain copies of the prescribed substantive and procedural standards if they exist?

E. WHEAT PRODUCTION AND DISTRIBUTION

1. How much land is sown to wheat in China? What proportion of the total sown land is this area? Have these magnitudes changed in recent years?

2. How much wheat is being produced in China? What proportion of total grain output is this quantity? Have these magnitudes changed in recent years?

3. How is the total output of wheat disposed of? What are the shares going to seed, feed, storage, urban consumption, and rural consumption?

4. How important are wheat imports to China and for what reasons? Does China plan to achieve self-sufficiency in wheat production, and if so, by what date?

5. How does China plan the production, distribution, and disposal of wheat? How are these plans implemented?

Mr. Hsü addressed himself to these questions in a meeting on June 15, 1976. He discussed questions A.1-A.4, B.2, and C.1-C.5. In addition, Mr. Hsü stated in response to question B.1 that TCK does not occur in China. He further noted that he could not answer questions D.1-D.5, because they are problems of other departments, and questions E.1-E.5, because the Chinese government does not publish data on crops.

APPENDIX F

WHEAT STUDIES DELEGATION LECTURE TOPICS

VIRGIL A. JOHNSON, *Chairman*

- Genetic Improvement of Wheat Protein
- Organization of Wheat Improvement Research in the United States

ROBERT H. BUSCH

- Evaluation of Early Generation Testing Applied to Adapted Spring Wheat Crosses
- Heterosis, Inbreeding, and Selfed Line Performance of Spring Wheat Crosses

R. JAMES COOK

- Soil-borne Diseases of Wheat and Their Relation to Cultural Practices and the Environment
- Biological Control of the Take-all Disease of Wheat

WARREN E. KRONSTAD

- The International Winter × Spring Wheat Hybridization Program
- Breeding for Maximum Grain Yield in Wheat

DALE N. MOSS

- Photosynthesis in Wheat and Barley
- Designing Wheat and Barley Plants for High Yield

ROBERT A. OLSON

- The Effective Use of Fertilizers in Wheat Production
- Commercial Fertilizer Use in Relation to Food, Energy, and Environmental Requirements

YESHAJAHU POMERANZ

- From Wheat to Bread
- Structure, Composition, and End-Use Properties of Cereal Grains

ALAN P. ROELFS

- Identification of Physiological Races of Wheat Stem Rust (*Puccinia graminis*)
- Methods Used to Control Wheat Stem Rust in the Great Plains of the United States

APPENDIX G

GERM PLASM PRESENTED

Materials presented to the Chinese Association of Agriculture were as follows (six packets each):

FROM OREGON STATE UNIVERSITY

16 winter wheats
6 spring wheats
16 spring triticales
10 spring × winter lines
11 spring × winter F_2 bulks

FROM THE AGRICULTURAL RESEARCH SERVICE (ARS) AND UNIVERSITY OF NEBRASKA

27 hard red winter wheat varieties
30 winter varieties from Eighth International Winter Wheat Performance Nursery
24 high protein and/or high lysine spring wheat lines
7 high protein and/or high lysine winter wheat lines

FROM NORTH DAKOTA STATE UNIVERSITY

16 spring wheats with semidwarf height, photoperiod insensitivity, early maturity, or high protein

FROM THE ARS RUST LABORATORY, ST. PAUL, MINNESOTA

9 spring and winter wheats with resistance to stem rust, leaf rust, or other wheat pests

FROM OREGON STATE UNIVERSITY

Spring Triticales

1. Beagle
2. UC 8825
3. Badger
4. M2A-1
5. M2A
6. S-603
7. Cinnamon
8. Inia/ARMADILLO
9. M^2A^2
10. Yoreme
11. Rahum
12. S-653
13. Beaver
14. Octo-Hexaploid
15. Koala
16. Octoploid Selection

Winter Wheats

17. McDermid
18. Yamhill
19. Hyslop
20. Tam 101
21. Roussalka
22. Sprague
23. Kirac 66
24. Sava
25. Bolal
26. Sturdy
27. Pregordnaia 2
28. Lovrim 13
29. Lovrim 11
30. Cleo/Pch
31. Edch/Mex 57
32. Dacia

Spring Wheats

33. Cajame
34. Wish
35. Borah
36. Anza
37. Yecora 70
38. Siette Cerros

Spring × Winter Wheat

39. JAR 66//CD*2/WW
40. P101/ANZA
41. DIBO//HQ/NA1
42. COFN/BEZO//DIBO/LFN
43. FN/3*TH//K58/N/3/MY54/N1OB//AN S/ANZA
44. TOB/YMH
45. SUW 92/Cl 13645//LFN
46. YMH/JAR
47. VIL 29/VG 59-8881//INIA'S'/DRC
48. COFN S/OFN S

Spring × Winter Wheat F_2 Bulks

49. SWM 741767 BEZO/2B//PERICO
50. SWM 741777 LFN/6/BB/CNO/5/LR64/3/SON//SKE/ANE/4/CNO
51. SWM 741858 MD/VG 9108//MD/OLOT/3/CNO'S'/PJ62//GALLO
52. SWM 741915 SON 64/2*SS/5/CC/KAL/4/A2//NAD/LR 64/3/BB
53. SWM 741940 SON 64/2*SS//CHANATE-2
54. SWM 742073 ROUSSALKA/3/TOB 66/B MAN//BB
55. SWM 742432 CD/PCH/3/NO 66/WS 1657//KAL/BB
56. SWM 742639 FORTUNATA*2/CI 13170/4/CHR/BB/3/CNO S/CAL/NAD
57. SWM 742754 CAPITOLE/PICH S
58. SWM 742808 BEZO/GALLO
59. SWM 742919 RIEB 4751/3/CNO S/PJ62//GALLO

FROM THE ARS AND UNIVERSITY OF NEBRASKA

Twenty-Seven American Hard Red Winter Wheat Varieties

Variety	Cereal Accession Number	Variety	Cereal Accession Number
Baca	15891	Parker	13285
Caprock	14516	Pawnee	11669
Centurk	15075	Sage	17277
Cheyenne	8885	Scout 66	13996
Comanche	11673	Scoutland	14075
Concho	12517	Sturdy	13684
Eagle	15068	Tascosa	13023
Gage	13532	Trader	13998
Hume	13526	Trapper	13999
Kharkof	1442	Triumph	12132
Lancer	13547	Wichita	11952
Lancota	17389	Winalta	13670
Minter	12138	Winoka	14000
Omaha	13015		

Thirty Cultivars in the Eighth International Winter Wheat Performance Nursery and Their Countries of Origin

Entry No.	Cultivar	Origin
1	Maris Templar	England
2	Maris Huntsman	England
3	Martonvasar 2	Hungary
4	Dunav-1	Yugoslavia
5	Flavio	Italy
6	GKF-2	Hungary
7	Biserka	Yugoslavia
8	TRS 237	Australia
9	Lely	Netherlands
10	Sentinel	Nebraska
11	Rashid	Iran
12	Kormoran	Germany
13	Kitakomi-Komugi	Japan
14	Talent	France
15	Bezostaya 1	USSR
16	Lerma Rojo 64	Mexico
17	Bordenave Puan Sag	Argentina
18	Galiafen	Chile
19	GKF-8001	Hungary
20	Martonvasar 3	Hungary
21	Oasis	Indiana
22	WWP 7147	Austria
23	Sage	Kansas
24	Odesskaya 51	USSR
25	Priboy	USSR
26	F26-70	Romania
27	WA 5829	Washington
28	NE 68719	Nebraska
29	Atlas 66	North Carolina
30	Blueboy	North Carolina

Spring Wheats from ARS-Nebraska Program with High Protein and/or High Lysine Potential for Breeding Purposes

1975 Yuma No.	Pedigree	Protein, %	Lysine, % of Protein		1975 Yield at Yuma, Ariz., kg/ha
			Unadjusted	Adjusted[a]	
10,038	Nap Hal/Atlas 66, F_7	16.0	3.0	3.2	4,400
10,100	Nap Hal/Atlas 66, F_7	16.0	3.1	3.3	3,700
10,121	Nap Hal/Atlas 66, F_7	17.0	3.0	3.2	5,900
	Comparative Performance of Parents and Check Varieties				
	Nap Hal	13.0	3.3	3.3	2,900
	Atlas 66	14.4	3.0	3.1	6,300
	Centurk	10.5	3.1	2.9	5,800
10,461	Nap Hal/C.I.13449, F_6	12.5	3.5	3.6	5,900
10,491	Nap Hal/C.I.13449, F_6	12.4	3.6	3.6	5,200
10,699	Nap Hal/C.I.13449, F_6	12.4	3.6	3.6	5,100
10,841	Nap Hal/C.I.13449, F_6	12.7	3.5	3.6	8,200
10,451	Nap Hal/C.I.13449, F_6	10.9	3.8	3.7	5,600
10,642	Nap Hal/C.I.13449, F_6	11.0	3.9	3.8	6,200
10,811	Nap Hal/C.I.13449, F_6	11.8	3.8	3.7	5,200

Comparative Performance of Parents and Check Varieties

	Nap Hal	13.0	3.3	3.3	2,500
	C.I.13449	10.2	3.5	3.2	6,200
	Centurk	10.1	3.2	2.9	4,600
10,958	Nap Hal/CIMMYT 8156	13.0	3.3	3.4	6,100
10,961	Nap Hal/CIMMYT 8156	16.9	3.1	3.3	7,100
10,967	Nap Hal/CIMMYT 8156	13.4	3.2	3.3	6,100
11,029	Nap Hal/CIMMYT 8156	12.3	3.4	3.4	6,900
11,074	Nap Hal/CIMMYT 8156	12.6	3.3	3.3	6,800
11,507	Nap Hal/CIMMYT 8156	13.6	3.3	3.4	7,500
11,511	Nap Hal/CIMMYT 8156	12.2	3.3	3.3	6,700
11,513	Nap Hal/CIMMYT 8156	13.1	3.3	3.4	6,300
11,522	Nap Hal/CIMMYT 8156	12.1	3.3	3.3	6,600
11,524	Nap Hal/CIMMYT 8156	12.9	3.3	3.4	6,400

Comparative Performance of Parents and Check Varieties

	Nap Hal	13.0	3.3	3.3	4,200
	CIMMYT 8156	10.2	3.3	3.0	5,900
	Centurk	10.7	3.3	3.1	4,000
11,564	Nap Hal/CB113	12.9		3.3	5,800
11,574	Nap Hal/CB113	13.0		3.3	6,200
11,593	Nap Hal/CB113	14.1		3.4	4,500
11,847	Nap Hal/Justin	13.1		3.4	4,000

[a]Lysine value adjusted to constant protein level.

Winter and Intermediate-Type Wheats from the ARS-Nebraska Program with High Protein and/or High Lysine Potential for Breeding Purposes

1975 Yuma No.	Pedigree	Protein, %	Lysine, % of Protein Unadjusted	Lysine, % of Protein Adjusted[a]	1975 Yield at Yuma, Ariz., kg/ha
21,061	C.I.13449/Centurk	12.6	3.2	3.2	8,500
10,486	Nap Hal/C.I.13449, F_6	12.9	3.6	3.6	5,500
10,521	Nap Hal/C.I.13449, F_6	12.5	3.5	3.5	5,900
10,661	Nap Hal/C.I.13449, F_6	13.0	3.4	3.5	7,700
10,768	Nap Hal/C.I.13449, F_6	12.0	3.5	3.5	6,600
10,185	Nap Hal/Atlas 66, F_7	16.2	3.1	3.3	4,800
10,600	Lancota	13.1	3.1	3.1	8,400
	Comparative Performance of Parents and Check Varieties				
	Nap Hal	13.0	3.3	3.3	2,500
	C.I.13449	10.2	3.5	3.2	6,200
	Centurk	10.1	3.2	2.9	4,600

[a] Lysine value adjusted to constant protein level.

FROM NORTH DAKOTA STATE UNIVERSITY

CHRIS: Insensitive to photoperiod. High protein. Origin, Minnesota.

ERA: Semidwarf height. Origin, Minnesota.

KITT: Semidwarf height. High protein. Origin, Minnesota.

NEEPAWA: High protein. Origin, Manitoba, Canada.

GLENLEA: Origin, Manitoba, Canada.

WALDRON: High protein. Origin, North Dakota.

OLAF: Semidwarf height. High protein. Origin, North Dakota.

ND 519: Insensitive to photoperiod. Early maturing. Origin, North Dakota.

ND 522: Semidwarf height. Origin, North Dakota.

ELLAR: High protein. Origin, North Dakota.

GWO 1809: Semidwarf height. Insensitive to photoperiod. Early maturing. High protein. Origin, World Seeds, Inc., United States.

PROFIT 75: Semidwarf height. Insensitive to photoperiod. Early maturing. Origin, World Seeds, Inc., United States.

PRODAX: Semidwarf height. Insensitive to photoperiod. Early maturing. Origin, Northrup King, United States.

PROTOR: Semidwarf height. Insensitive to photoperiod. Early maturing. Origin, Northrup King, United States.

BOUNTY 208: Semidwarf height. Insensitive to photoperiod. Early maturing. Origin, Cargill, Inc., United States.

NOWESTA: High protein. Origin, North Dakota, private.

FROM THE ARS RUST LABORATORY, ST. PAUL, MINNESOTA

McNAIR 701: C.I. 15288. An early, short-strawed, brown chaff, soft, red winter wheat. Has resistance to Hessian fly *Mayetiola destructor* (Say), powdery mildew (*Erysiphe graminis* D.C.), and leaf rust (*Puccinia recondita*) (*LR*-9). It is too susceptible to *Septoria nodorum* (Berk) and stem rust (*Puccinia graminis*) to be recommended for commercial production. Will head without much vernalization.

AGENT: C.I. 13523. An *Agropyron elongatum* derivative. This hard, red winter wheat has leaf rust resistance (*LR*-24) and stem rust resistance (*Sr* 24). Will head without much vernalization.

AGATHA: C.I. 14048. An *Agropyron elongatum* derivative in a hard, red spring wheat with leaf and stem rust resistance (*LR*-25, *Sr* 25). Has not been used commercially owing to a yellow gluten color.

WRT 238-5: C.I. 14141. A spring wheat with stem rust resistance from Imperial Rye (*Sr* 27). Not grown commercially.

W2691SrTt-2: A semiclub wheat with *Triticum timopheevi* stem rust resistance in a hexaploid wheat. Not used commercially.

IDAED 59: C.I. 13631. A white spring wheat with a gene from *Triticum timopheevi* (*Sr* Tt-1) for stem rust resistance. This gene is associated with a low rust severity in the field. Some leaf rust, stripe rust, and powdery mildew resistance. A commercial variety in Idaho.

OASIS: C.I. 15929. A soft, red winter wheat with resistance to Hessian fly, leaf rust (*LR*-9), powdery mildew (*Triticum timopheevi*), stem rust (*Sr* Tt-1), and *Septoria tritici* (Bulgaria 88).

STURDY: C.I. 13684. A semidwarf, hard, red winter wheat with adult plant resistance to leaf rust. A commercial variety in Texas.

SWSr22T.B: A hexaploid wheat with stem rust resistance from *Triticum boeticum*. Not grown commercially.

APPENDIX H

GERM PLASM RECEIVED

The Chinese Agriculture Association presented to the U.S. Wheat Studies Delegation three samples each of 48 winter wheat varieties, 22 spring wheat varieties, and two triticale varieties, for a total of 216 samples:

USDA Plant Intro- duction No.	Packet No.	Varietal Name	English Translation
Winter Wheat Varieties			
414593	1	Nung ta 311	Agricultural College 311
414564	2	Tung fang hung 2	East is Red
414590	3	Nung ta 139	
414596	4	Peking 9	
414621	5	Yang mai 1	
414565	6	Tung fang hung 3	
414591	7	Nung ta 141	
414568	8	Feng ch'an 3	High Yield
414617	9	Shuchou 14	From Shuchou
414595	10	Peking 8	
414556	11	Ai feng 3	Dwarf High Yield
414567	12	Ao mai 6 (Ngo mai 6)	From Hupei
414599	13	Peking 15	
414580	14	Hung liang 10	Red Lady
414594	15	Peking 6	
414608	16	Shihchiachuang 34	From Shihchiachuang
414598	17	Peking 11	
414565	18	Ai kan tsao	Dwarf Early
414559	19	Pai yu pao	White Big Grain
414597	20	Peking 10	
414592	21	Nung ta 147	
414558	22	Anhwei 11	From Anhwei
414605	23	Shih 3626	
414602	24	Shih 70-17	
414607	25	Shih 4468	
414603	26	Shih 70-22	
414606	27	Shih 4182	
414622	28	Chi mai 23	From Hopei

USDA Plant Introduction No.	Packet No.	Varietal Name	English Translation
414610	29	Shih p'in 83	From Shihpin
414577	30	Ke hsuan 1	New Selection
414578	31	Ke hsuan 2	New Selection
414613	32	Hsiang yang 1	Phototropic
414614	33	Hsiang yang 2	Phototropic
414615	34	Hsiang yang 3	Phototropic
414609	35	Shih p'in 10	
414604	36	Shih chung 14	
414557	37	Ai feng 4	Dwarf High Yield
414579	38	Kuan ts'un 1	From Kuantsun
414612	39	Wu nung 132	From Wuhan
414619	40	Yenan 6	From Yenan
414581	41	Chin ch'ung	Bright Light
414601	42	Shan nung 6521	From Shansi
414625	43	3311	
414587	44	Ching yang 302	From Chingyang
414620	45	Yenan 15	From Yenan
414618	46	Ya chiao hung	Duck Foot Red
414624	47	Ku tzu t'ou	Drill Head
414582	48	Ching ai tsao shu 21	Dwarf Early from Hupei
Spring Wheat Varieties			
414588	1	Ke feng 1	High Yield
414583	2	Ching hung 1	Peking Red
414584	3	Ching hung 2	Peking Red
414616	4	Hsin shu ch'ung 1	New Gloom
414623	5	Yu yi mai	Friendship
414586	6	Ching hung 6	Peking Red
414585	7	Ching hung 3	Peking Red
414560	8	Ts'ao yuan 1	Grassland
414561	9	Ts'ao yuan 2	Grassland
414562	10	Ts'ao yuan 3	Grassland
414563	11	Ts'ao yuan 4	Grassland
414569	12	Feng ch'iang 1	High Yield, Strong Stand
414570	13	Feng ch'iang 2	High Yield, Strong Stand
414600	14	Ch'ing ch'un 5	Youth
414571	15	Kung chiao 277	Common Bread
414572	16	Kung chiao 279	Common Bread
414573	17	Kung chiao 282	Common Bread
414574	18	Kung chiao 284	Common Bread
414575	19	Kung chiao 287	Common Bread
414576	20	Kung chiao 288	Common Bread
414611	21	Shan nung 17-17	From Shansi
414589	22	Ke tsao 7	Drought Resistant
Triticale Varieties			
	1	Triticale 2	
	2	Triticale 3	

APPENDIX I

PUBLICATIONS PRESENTED

Agricultural Research Service. *Wheat in the United States*. Agricultural Information Bulletin 386. Washington, D.C.: U.S. Department of Agriculture. (18 copies)

Baker, K. F., and R. J. Cook. *Biological control of plant pathogens*. San Francisco: W. H. Freeman and Co. (6 copies)

Foreign Agricultural Service, Quality Wheat for the World (handout). Washington, D.C.: U.S. Department of Agriculture. (10 folders)

Olson, R. A., *et al.*, eds. *Fertilizer technology and use*. Madison, Wisc.: Soil Science Society of America. (4 copies)

Proceedings of the Second International Winter Wheat Conference, Zagreb, Yugoslavia, June 9-19, 1975. (6 copies)

APPENDIX J

PUBLICATIONS RECEIVED

Acta Botanica Sinica

Vol. 15	No. 1	1973	(2 copies)
Vol. 15	No. 2	1973	(1 copy)
Vol. 16	No. 1-4	1974	(3 copies)
Vol. 17	No. 1-4	1975	(3 copies)
Vol. 18	No. 1	1976	(2 copies)

Acta Genetica Sinica

Vol. 1	No. 1-2	1974	(2 copies)
Vol. 2	No. 1-4	1975	(2 copies)
Vol. 3	No. 1	1976	(2 copies)

Acta Phytotaxonomica Sinica

Vol. 12	No. 1-4	1974	(2 copies)
Vol. 13	No. 1-4	1975	(2 copies)

Chung-kuo nung tso wu ping ch'ung t'u tsao. Part 2. Mai lei ping chung (Wheat diseases and insects). 1974. 2nd ed. Peking: Agriculture Press. (2 copies)

Institute of Botany of the Chinese Academy of Sciences, comp. 1972. *Iconographia cormophytorum sinicorum.* 4 vols. Peking: Science Press. (1 set)

Peking Academy of Agricultural Sciences, comp. 1975. *Mai t'ao yü-mi tsai-p'ei kuan-li li-ch'eng* (Management of cultivating corn interplanted with wheat). Peking: People's Press. (2 copies)

Peking Wheat Cooperative Group, comp. 1975. *Pei-ching shih hsiao-mai kao ch'an ching-yen hsüan-pien* (Selected experiences with high yields in wheat in Peking). Peking: People's Press. (2 copies)

Shih, S. H. 1974. *A preliminary survey of the book "Ch'i min yao shu": An agricultural encyclopedia of the sixth century.* 2nd ed. Peking: Science Press. (2 copies)

Shih, S. H. 1974. *On "Fan sheng-chih shu": An agriculturist book of China written in the first century B.C.* 3rd printing. Peking: Science Press. (2 copies)

Zhongguo Nongye Kexue, 1976. No. 1. (2 copies)

Reprints of several articles published by members of the Peking Institute of Botany of the Chinese Academy of Sciences.

Reprints of several articles published by members of the Shanghai Academy of Agricultural Sciences.

APPENDIX K

A CONVERSION TABLE OF WEIGHTS AND MEASURES

AREA

1 *mu*	= 6.666 ares	= 0.164 acres
1 acre	= 0.4046 hectares	= 6.070 *mu*
1 hectare	= 15 *mu*	= 2.471 acres

WEIGHT

1 *jin*	= 0.5 kilogram	= 1.1023 pounds
1 kilogram	= 2 *jin*	= 2.2046 pounds

LENGTH

1 *li*	= 0.5 kilometer	= 0.310 mile
1 mile	= 1.6093 kilometers	= 3.218 *li*
1 English foot	= 0.3048 meter	= 0.9144 Chinese feet

CAPACITY

1 U.S. gallon	= 3.7853 liters	= 3.7853 Chinese liters